Palgrave Studies in Sustainability, Environment and Macroeconomics

Series Editor

Ioana Negru, Department of Economics, SOAS, University of London, London, UK

Most macroeconomic theory and policy is orientated towards promoting economic growth without due consideration to natural resources, sustainable development or gender issues. Meanwhile, most economists consider environmental issues predominantly from a microeconomic perspective. This series is a novel and original attempt to bridge these two major gaps and pose questions such as: Is growth and sustainability compatible? Are there limits to growth? What kind of macroeconomic theories and policy are needed to green the economy?

Moving beyond the limits of the stock-flow consistent model, the series will contribute to understanding analytical and practical alternatives to the capitalist economy especially under the umbrella term of "degrowth". It will aim to reflect the diversity of the degrowth literature, opening up conceptual frameworks of economic alternatives – including feminist political ecology - as critical assessments of the capitalist growth economy from an interdisciplinary, pluricultural perspective.

The series invites monographs that take critical and holistic views of sustainability by exploring new grounds that bring together progressive political economists, on one hand, and ecological economists, on the other. It brings in.

More information about this series at
http://www.palgrave.com/gp/series/15612

Ion Pohoaţă · Delia Elena Diaconaşu ·
Vladimir Mihai Crupenschi

The Sustainable Development Theory: A Critical Approach, Volume 2

When Certainties Become Doubts

Ion Pohoaţă
Alexandru Ioan Cuza University
Iasi, Romania

Delia Elena Diaconaşu
Alexandru Ioan Cuza University
Iasi, Romania

Vladimir Mihai Crupenschi
Alexandru Ioan Cuza University
Iasi, Romania

ISSN 2635-2621 ISSN 2635-263X (electronic)
Palgrave Studies in Sustainability, Environment and Macroeconomics
ISBN 978-3-030-61321-1 ISBN 978-3-030-61322-8 (eBook)
https://doi.org/10.1007/978-3-030-61322-8

This Palgrave Macmillan imprint is published by the registered company Springer Nature Switzerland AG
The registered company address is: Gewerbestrasse 11, 6330 Cham, Switzerland

Introduction

This volume is the twin sibling of the first volume entitled *The Sustainable Development Theory: A Critical Approach, The Discourse of the Founders*.

As the subtitle tries to convey, in the following pages we would like to find out why and how many great certainties of the classical and neoclassical founders, pioneers in the area of sustainable development, are being cast by modern and contemporary authors as doubtful or uncertain. In fact, this is the message of our book. We are trying to provide circumstances and arguments to prove that the painful and unforgivably strong detachment from the core ideas of the great founders does no good to either the theory or the practice of sustainability.

On all three levels—economic, social and environmental—the founders' works remain worth quotable. The "productive trinity" had its methodological duplicate. They did not separate social and environmental problems from the pure exercise of material production. Synthesis seemed to them the natural path to take. Therefore, far from being blatant and emphatic, but in a dialectical way, they understood things, down to their intimate organic structure. The social embroidery that accompanies Adam's work equally invites living in harmony with nature. However, the "eggs" they laid have not always hatched worthy chickens. The Brundtland Report itself separated them from the "nest". The disjunction on the three mentioned levels and the emergence of distinct versions of some environment-related economic sciences (regardless of their name) reflect such a centrifugal trend. Mill's stationary state repackaged as "zero growth" or "degrowth" goes down the same path. The economists' divide into economists as such and environmentalists is also a case exemplifying this disintegrating exercise. The effort made in such a direction might seem justified, worthwhile. Yet, we see it as a regrettable detachment from the whole. The advantages of a beneficial process of division of labour do not compensate for the risk of having the economic science classified as an alternative to substitutes maintained by scholarly controversies directed in artificially created contexts, nor do they cover the necessary recrystallization efforts.

The intellectual exercise of the great founders was both an integrator and a lesson of objective honesty, intellectual refinement and thought rationalization. They presented the world with no run-arounds. There was no hypocrisy in it; it was all critical ability. They considered that all the big questions about the functioning of the economy and society are worth asking. Even when they addressed the problems of slavery or the thorny issue of population dynamics. Or, why not, when they realized that those who toil—with their hands or heads—are men and women, engaged in effort according to the principles of a natural division of labour in which specificity, natural or societal differentiation can be non-discriminatory and non-disturbing; on the contrary, they may have a harmonizing, integrative potential. In addition, even when, with visible but justified infatuation, they thought they were universal creditors of ideas, they were guided by a special responsibility. As established scholars, they felt empowered to instil, in addition to rational behavioural and judgement principles, a social principle, closely linked to ethics and morality. Their world acquired the attributes of sustainability through their manifest concern to push the principles of rationality beyond the strict boundaries of economics; they took a strong interest in ethics and morality, education and culture, and in social, environmental, political and cultural issues. Their world, we repeat, could only be civilized through education. Only an educated population could acquire that "ability", a key word in the Brundtland Report, and, hence, achieve prosperity, in a continuous, resilient way, as we say nowadays, in accordance with the social equilibrium and the harmony of nature. Even if they did not name it, they had a very clear image of what a sustainable society should look like. They did not capture its essence in sophisticated models or in writings burdened with statistical data or attempts to validate farfetched hypotheses. Besides, this was not even necessary. Life was simpler and more sincere. Even the banks were tempted by the purifying principle of the natural rate and by the combination of their activity with that of production, which they used as a nurturing source. Supreme happiness could not be acquired materially; even more so, the immaterial component had a major say. The environment itself was generous and unfettered. Inspired by such a landscape, the founders produced ideas pertaining to sustainability. Not sustainability hidden behind curtains, but visible, inspiring, worth quoting and mimetic.

The analysis offered in this volume tries to prove that indeed, in terms of sustainability, the founders are worth quoting, and if their pathway is abandoned, truncated or insidiously interpreted, sustainability stands to lose. Engaged in such an exercise, we cannot deviate from their methodology. We found inspiring not only their scientific audacity but also their bravery and honesty. As such, we took into account the authorities without letting ourselves be overwhelmed by celebrated names. We insisted on the interrogative, critical, objective and anti-canonical dialogue, despising compromise.

If in the first volume, we set out to decipher texts and messages to find out what the founders thought about sustainability in their own terms, in this volume, with reference to the aforementioned mirror, we seek to find out if the current theory of sustainable development still preserves, and to what extent, the founding dowry. First, our work involved selection. We logically deemed that the structure of the book allows for comparison, as much as possible in terms of time and ideational content. We looked for themes and authors that offered compatibility and comparability; what existed, what was conceived by thought and what has been transmitted in current theoretical bodies. We looked for powerful ideas, selecting economists who had something to say and convey to the next generations, passing over the reed and insistently sterile squeak that currently accompanies the topic.

In the previous volume, a first chapter was dedicated to explaining the operable notional system and the epistemological base on which the conceptual structure is built. We recall here that we relied on the definition relegated to the status of message by the Brundtland Report. A message which encourages economic development by acquiring "ability", in harmony with nature and for the individual and collective wellbeing of people; a message applicable to a diverse world, both as a project and as a result. The theoretical-doctrinal developments featured in the first volume focus on the Brundtland Report, with the two working concepts—sustainability and durability—synonymous in meaning but different in form, depending on the theoretical framework involved, of English, or French origin. Concerned with the durability—sustainability—resilience kit, we concluded that "(…) the simple assertion that an economy is sustainable means that its dynamic comprises robustness, viability and a capacity to overcome social, environmental or resource constraints". If those who read the present volume find that sustainability is, at times, replaced as a working phrase by sustainable development, this should not affect the whole, because it does not change the quality of the analysis, the meaning of ideas and the accuracy of the book's message. We have only short-circuited the wording and let the context speak for itself.

The second volume is accurately read and understood if naturally seen as an extension of the first volume. It is there that we posed the problem. Here we are trying to see how it is solved by contemporaries. Looking at some selected models of analysis, just to pinpoint the progress (or the downgrade), intrinsically turns the text more technical, forcing digressions on large areas of the economic theory, about sustainability and not only. To those who are tempted to "dodge", to skip the first volume, although not desirable, we offer a foothold. We reiterate some ideas, both to provide support and to entice the reader to access the sources.

Thus, consistent with the founders' thought, we considered and argued that inequality and harmony are not irreconcilable. Nature places us in different positions at the starting line, and we remain just as different after school and society do their part. We deemed void the claim to sustainability starting from the premise of a society populated by homogeneous human units. The thin distinction between people remains a natural and acceptable process. Thomas Friedman's "flat world" is populated by different individuals. The division of labour among educated or less educated people remains fundamental to the process of development, and human cooperation is the pendant of this division.

The social harmony of the founders rejects mediocrity. Everyone must be "well fed, dressed and sheltered" but the "salt of the earth", the elite, must be allowed to flow like Niagara Falls. The family and the state remain the main pillars responsible for education. The problem of the intergenerational relationship is interconnected with the family policy and it captures the "natural" endemic vein in shaping the feeling of societal care.

With different people and rules, the founders' development model is mosaic framed. You cannot copy it, on the whole. The founders provided us with solid arguments to support the fact that the foundation of a world with the attributes of sustainability is only possible with the contribution of a hard core of rules and good practices; in other words, it needs institutions. The first volume presents the anatomy of the common ground of ideas of classical institutionalism, as well as the circumstances that render J. S. Mill a special institutionalist. Even if the sustainable development project is relatively young, the "old" institutional offer can foster useful insights into the importance of the "intelligence of rules".

The founding lesson on economic dynamics remains intact. Some key phrases may be revealing: (1) It is healthier to invest from savings rather than going into debt; (2) Accumulation has been a revolutionary idea; (3) Capital and labour share an intimate relationship of complementarity; (4) There is a sequential, canonical order of dynamics, in which the tone is set by production; (5) Economy has its self-regulatory mechanisms; that is, there is self-development and the promotion of growth based on either demand or supply, as two distinct entities, is a false and unsustainable idea; (6) Economic dynamics guarantees social dynamics. That is, dynamics is desirable, not without limits, but necessary and the steady-state is only admitted as a well-deserved break following a productive effort; (7) The management of economic dynamics is purely national. A country can only be imperfectly ruled by foreigners.

Looking at some of the founding neoclassicists, we learned in the first volume that the Brundtland philosophy rejects the Walraso-Paretian optimum and equilibrium. The way in which Menger's picture reconciles with the rational and wealthy man, and the way in which it is incompatible with the diversity of human preferences and the frail demand of the poor man, is another worthwhile idea of the volume. The neoclassicists also take interest in analyses of absolute and relative poverty; the reasons why the right to inheritance should be reassessed; the role of the institution of entrepreneurship as seen by Marshall or Menger; Pigou's externalities or Jevons' rebound effect. The glue that binds together the social dimension of a sustainable development in the formula welfare—social justice—distribution profusely draws attention.

In regard to what competition can offer, as an exploitable subject in support of sustainability, the first volume explores less surveyed areas. The analytical pathway opened by the natural price of goods and money, as initiated by the founders, is a generously exploitable area in support of sustainability. The link between generations through interest and the roundabout production technique, as well as the Wicksellian synthesis of the natural interest rate, including the anatomy and physiology of the cumulative process—these all are just a few tempting reasons for the eyes and minds of those interested in the subject to read the first volume as well. We hope they will, and that they will be as numerous as possible. Looking back, as we also did in the first volume, hoping to discover the enhancing bounty provided by the inherited dowry we rely on today, we noticed in the present volume two aspects. First, at least in the formal manner, today almost everyone builds on his own. Although the long-range ideas of the founders configure, through a natural cumulative process, the architectural structure of the mainstream, the citation, and the direct references to them are no longer in vogue. Most of the founders seem to be forgotten. The second observation—an extension of the first—is that the speech of the descendants shows traces of the ballot boxes—either in support of some master that common sense urges to follow, or for some doctrine whose system is confining thoughts. The conclusion? The message of the descendants is, in broad lines, indefinite and equivocal. Looking for a coherent and cohesive lesson for sustainability, we discovered a great deal of labyrinths, Babel Tower areas, quarrels between idealistic colleagues and, mostly, a lot of grief as far as the conclusions are concerned. We hope that the pages to follow will shed light on these thoughts.

Contents

1 First Steps in Perverting Sustainability. Keynes—A Good Omen? 1
1.1 A Founding Obsession on the Path to Sustainability 2
1.1.1 The First Step Towards Unsustainability! Menger: Money Does Not Measure Anything! 2
1.1.2 Mises Versus Keynes at the Crossroad of Money That Does Not Measure Anything 6
1.2 Sustainability of Anti-crisis Therapy: Keynes Versus Hayek 12
1.2.1 Keynes: A Revelation and an Unclogged Road for the Problem of Sustainable Business Cycle 12
1.2.2 Hayek: When the Sustainability of the Business Cycle Rests on Shaky Premises 17
1.3 Concluding Remarks 40
References 42

2 In Search of a Lost Lesson 45
2.1 Exploring the Outskirts of the Business Cycle Sustainability 45
2.1.1 "Heresies" of the New Austrians on the Topic of Business Cycle Sustainability 46
2.1.2 On the Sustainability of the "Monetarist" Cycle 50

2.1.3 Irving Fisher: Over-Indebtedness—Deflation—Monetary Injection. And if You Are Poor, Chances Are You Remain So! 54
2.2 Renowned Interpreters in Search of a Sustainability Deprived of Real Sources 56
2.2.1 Mason and Butos: Keynesianism Is Sustainable, Monetarism Not at All! 57
2.2.2 The "Minsky Moment": An Erudite Analysis Lacking Sustainable Message 60
2.3 Can We Find Sustainable Ideas at the 2008 Crisis Theorists? 66
2.4 Concluding Remarks 68
References 71

3 How to Conceive the Brundtland Agenda in the Context of the Nominal Economy's Imperialism 75
3.1 How Would an Economic and Sustainable Up-to-Date Agenda Look like If Conceived According to the Bruntland Matrix? 76
3.2 Three Delicate Areas Where the Recourse to the Founders Would Lead the Way Towards Sustainability and Resilience 77
3.2.1 Playing with Money When Dealing with Inflation and Distribution 78
3.2.2 Masters of the Economy. From Instrument-Actors to Masters that Set the Rules 81
3.2.3 Achieving Sustainability Through the Natural Rate of Interest—A Necessary Synthesis of Ideas 86
3.3 Post-Wicksell and Postcrisis. Current Metamorphoses of the Natural Rate of Interest 90
3.4 Instead of Conclusions: Why Does the Central Bank Refuse the Sustainability of the Natural Rate of Interest? 97
References 100

4 Social Pressure—The Risk of Invalidating the Lesson on What Makes an Economy Sustainable 103
4.1 Can Social Tensions Shift Causalities and Undermine the Logic of Sustainability? 103
4.2 Economism and Anti-Economism—Ideologies at the Borders of Sustainability 109
4.2.1 A Brief History 109
4.2.2 New Characteristics of Anti-Economism 111
4.2.3 GDP Statistics—Between Economism and Anti-Economism 114
4.2.4 Who Criticized and Still Critiques Profit? 118
4.3 From Anti-Economism to the Court of Distributive Justice 122
4.3.1 From Arithmomorphism Towards Dialectics and Distributive Justice 122
4.3.2 Distribution Before Production. The Illusion of Distributive Justice 125
4.3.3 The Poison of Profit and Productivism 131
4.4 Pikettism—The Zeal of Quantitative Levelling 136
4.5 Concluding Remarks 153
References 158

5 Degrowth—A Logical Inadequacy? 163
5.1 The Source of the Confusion: Consolidating Scientific Props! 164
5.2 New Meanings in a New Language 168
5.3 Degrowth in Search of a Credible Fixture 170
5.4 Degrowth Between Yes and No 177
5.4.1 From the Puzzling Suggestion that Degrowth Is An Invitation to Stop by Continuing to Run 179
5.4.2 From the Contrast Between the Results Promised by Degrowth and the Utopian Character of the Project 182
5.4.3 Because of the Lack of Coherence in the Remedies for the Developing World 185
5.5 The Scenario-Manic 190
5.6 Concluding Remarks: Between the Natural Rate and 2° Celsius! 195
References 202

6 Nature – The Highlight of the Theory of Sustainability 205
6.1 The Dialectics of a Fundamental Relationship: MAN - NATURE - MACHINE 205
6.2 Substitution in the Vision of the Theorists 216
6.2.1 Strong Sustainabilists: Natural Capital is Indestructible! 217
6.2.2 The Solow Approach—self-Development Through Ability À La Brundtland 219
6.2.3 Can Sustainability Be Nourished from the Solow-Daly Querella? 223
6.3 Nature Versus Institutions. The Institutional "creative Destruction" 227
6.4 Marx's "Machines" and Contemporaneity 232
6.4.1 Substitution Supported by Perverse Dialectics 232
6.4.2 From Marx's Machines to Its Highness, the GENERAL INTELLECT. The Multifactor Productivity 235
6.4.3 Sustainability in the Era of the Technological Unemployed 240
6.5 Is There Need for a New Science to Achieve Strong Sustainability? 242
6.6 Sustainability and the World of Animals 251
6.7 Concluding Remarks 255
References 261

7 General Conclusions 267

Index 281

CHAPTER 1

First Steps in Perverting Sustainability. Keynes—A Good Omen?

Human action has long ceased to be a mere human-nature equation. On the contrary, it is mediated, mixed, delegated, complex, contradictory and abounding in both the rational and the irrational; in deception and fetishism. Among all, the fetishism of money clearly stands out. Ever since it was called to mediate exchanges—at minimal cost—up to the point of reflecting real wealth, money has filled the economic science with mysteries, areas "forbidden" to the common mind, or with crucial problems—unsolved or in the process of being solved.

Although it complicates it, money never leaves the world of tangible goods and services. The economy, as a phenomenon, reality, and science of the material wealth, coexists with this reflection. Over time, the reflection shows signs that it may be more important than the reflected object. Development theory cannot bypass this subject. The alleged autonomy of the nominal gives room to put forward situations likely to trigger the great crises that both the real economy and the ones that it has to feed, dress, shelter, etc., suffers from. It would be ideal to surgically eradicate the eventual illnesses of the nominal from their root. However, when money is the blood of the real organism, the operation is no longer possible. The nominal-real separation becomes a mere logical inadequacy. That is, we need to ensure the health and sustainability of the whole.

I. Pohoaţă et al., *The Sustainable Development Theory: A Critical Approach, Volume 2*, Palgrave Studies in Sustainability, Environment and Macroeconomics, https://doi.org/10.1007/978-3-030-61322-8_1

A whole in which the nominal is ingrained—worked in its own organic texture.

There is no area of economics to have multiplied teaching technicality as much as nominal economics did. Just as no other exceeds it in terms of the diversity of viewpoints on the same subject. Failure to align to a unique point of view is likely to render it alive and interesting. Nevertheless, the doubt of some alleged truths—we dare say—do not render it as solid and sustainable. After all, how important are divergent opinions in the world of money circulation? Can they explain the pluses or minuses in the dynamics of an allegedly sustainable and resilient economy? Let us try to meet this challenge.

1.1 A Founding Obsession on the Path to Sustainability

Not all the ideas of the founders made it through time. This is also due to a well "justified" contemporary infatuation, according to which progress allows us to know more than our predecessors did. We tend to forget that we are how they wanted us to be, that is why they are the founders, those who have shaped us. One idea that is clearly overlooked in this exercise is related to money, and to its role and functions. It crosses, at a higher or a lower intensity, seen or bypassed, the whole history of economic analysis. We analysed it too because we consider it to be closely connected to the fundamental sources of sustainability; or, as the case may be, the lack of sustainability.

The contemporaneity of the idea is extremely complicated. Therefore, we need to select a representative sample, with all the inherent risks of such a procedure. We will thus focus on a couple of examples only, namely on a few landmarks: Menger, Mises, Hayek, Keynes. From them and through them, we will also touch upon others, no less important. Here are the ideas that reflect their name!

1.1.1 The First Step Towards Unsustainability! Menger: Money Does Not Measure Anything!

Why did money appear? Was it to measure values or to serve as a universal means of exchange? Or was it both? The problem seems overused, although it is not! It remains just as important today as it was when it was first put forward. Moreover, it is equally interesting today and here, for the purposes of our study, as the answer to this question accounts for many of the statements underlying the theory of interest rates, inflation and the economic cycle, with great impact on sustainability.

It is no coincidence that the idea split the classicists into two groups. Some, led by Say (1971) and Bastiat (2007), courted the subjective value and understood money's role as an exchange vehicle. Smith (1977) took the middle path, but Mill (1885) and, in particular, Ricardo (2001)—in their quest of an absolute measure of value—launched also the idea of money as potentially fulfilling the role of measuring value. They launched it and later left it. Marx (1990) followed in their footsteps and went all the way, completing this endeavour. Beyond the ideological shell in which he enclosed his idea, the Marxist analysis on the origin and functions of money remains a landmark.

With the marginalist revolution, objective value gives way to subjective value. Without demand, the cost of goods is useless. Value is an appreciation; it is subjective because it belongs to the subjective judgement of the individual and not to accounting calculations pertaining to the upstream of the free market. Marshall (2013) reconciles the "schools" by making it clear that "the two blades of a pair of scissors" are just as important; labour (cost) remains the source of value; however, its size remains to be assessed based on its utility. Nothing clearer. What about money, its origin, quantity, forms, and functions? Things are not as clear here!

Four years after the *Capital* came out, where in his first volume, Marx (1867/1990) sets out his point of view, Menger's (1871/2007) *Principles* was published. Relying on the same tone and arguments as Marx, this is where he demonstrates—as we have shown in the first volume, where we discussed about money as institution—that money was not invented; that it did not come out of nowhere, but was objectively called by the economy in order to make exchanges possible and at minimal cost. Just like Marx, Menger demonstrates that money imposes itself as a "shared belief", as an institution. But while the former sees "money as a measure of value" the main function from which all the others derive, for Menger, the function of "means of exchange" seems to be the real *raison d'être* of money. What is the difference? The difference is actually fundamental. That is why it has functioned as a starting point for those who tried to explain the "reason" of the quantitative surplus of money as a "safe" solution, one that is relaxing and possible in a major crisis.

Menger was obsessed with the problem but was unable to settle it. Constantly getting back to the issue—in an attempt to clarify it—his struggle to shed light on things had the opposite effect. After the *Principles*, where he talks about *The Theory of Money*—Chapter VIII, he resumes the discussion and writes three articles, driven by the same goal—to emphasize the fact that the function of money conceived as a means of exchange is instrumental and sufficient. One of the versions came out in French with the inciting and vexing title *La monnaie mesure de valeur* (Menger, 1892)/*Money as Measure of Value*. Despite what the title announces, Menger is looking for arguments to convince that it is not the dominant, fashionable doctrine (he was thinking of Marx although he does not mention his name) that should be taken into consideration. No, he believes value is not what the dominant theory states it is. It is subjective instead of objective; then, there is no reason why money should measure it; there is no need for measurement. He completely refutes the point of view according to which the main function of money is that of providing a measure of value. However, his demonstration is not as obvious and convincing. Following his work, it is not hard to notice the difficulties and the half-measures in his approach on the subject. Here is what he writes, towards the end, under the title *Money as Measure of Price*, in Chapter VI of the article: "The old theory lies upon the idea that *the equality of values is the main concern in the exchange process*. Neither one [partner] nor the other thinks in the least of exchanging some equal value for another equal value: *the goal they follow is to satisfy their needs* (…) Exchange does not require any previous measuring. (…) Money has become the intermediary of exchange, but if it serves well in measuring prices, it is only in the sense that we have just pointed out" (Campagnolo, 2005, p. 260). The end of the article "makes it clear" that price has nothing to do with any measure of value; it is a result of the negotiation between the two players, seller and buyer, where "each of them has played a part in setting that very price according to how they may benefit the most" (Campagnolo, 2005, p. 261). Menger signs this text and believes that he has finally clarified the problem. Yet, the faults in thought and judgement are too visible. If the exchange "does not require any previous measuring", how is it then possible to outline an exchange? If we do not use the concept of money, then we can say that 10 eggs = 1 book. In physical terms, subjective assessment is admissible. However, we do it relying on money, mentally, abstractly, without actually counting money. Let us nevertheless agree for a moment with Menger and consider that this step is passed. What happens after the two players make their assessment and agree that the exchange is mutually beneficial? They start talking about price, aren't they? Moreover, if the price is not the expression in money of this subjective value, then what is? Could Menger set out to conduct an exchange without assuming that money is useful? Without admitting that money is useful in assessing, in putting a price on value, in making a gradual and subjective estimation of value? And that it is precisely this power—inherent and granted by origin—that allows money to serve as a means of exchange? Or payment, saving, investment, etc.?

As already argued, the quoted article was intended as a backtrack to the analysis in *The Theory of Money*, Chapter VIII of the *Principles*; a deliberately enticing analysis, accompanied by mind-blowing conclusions. From the very beginning, it can be noticed that the author set out to unyieldingly deny the function of money as a measure of value. He set out to do so, but it did not come out as intended; the text is full of tautologies and vexatious omissions. The phrase money, considered as a "Measure of Price", is likely to irritate, by its mere statement, any economic logic. The statement according to which "the notion that attributes to money (…) the function of also transferring 'values' from the present into the future must be designated as erroneous" (Menger, 2007, p. 279) is an unsupported "commandment" given by a founder of the school. Weren't values passed on from one generation to the next at the time? In the subtitle "Money as a 'Measure of Price' and as the Most Economic Form for Storing Exchangeable Wealth" he admits that it is possible to store by means of and with the help of money. We do accumulate values, expressed in money, don't we, Menger implicitly admits, and we do so not in vain, but in order to pass them on to future generations. Self-consciousness follows him through, and thus, puts the expression—measure of price—between inverted commas.

The examples could go on, but they lead to the same conclusion. In short, Menger breaks away with Ricardo and especially Mill, Marx and their analysis of money. He follows Marx's argument, but his conclusions are "decreed". Instead of standing out from the causal line of argument, these conclusions take the form of doctrinal prohibitions. In other words, he urges: do not speak like them; like them, or like those who prefaced them. In fact, the statement that "money as the 'measure of the exchange value' of goods disintegrates into nothingness, since the basis of the theory is a fiction, an error" (Menger, 2007, p. 273) is a label also put on Turgot's analysis. Theories that are "untenable", he will ordain on page 276 referring to all those who saw things differently from him! At the embarrassingly apodictic and infatuated end of the chapter, we discover a Menger terrified of the idea of becoming infected—in this way—with the "Marxist leprosy" and crouched under the commands of the platoon commander. Here it is: "the functions of being a 'measure of value' and a 'store of value' must not be attributed to money as such, since these functions are of a merely accidental nature and are not an essential part of the concept of money" (Menger, 2007, p. 280). "We understand", all the disciples will say!

In relation to such a disqualifying end of the road, Menger felt the need to come back. He came back and made things even more confusing. In addition to the great ideas, such as the causal connections between things and the satisfaction of human needs, the complementarity of consumers' goods and production factors, the costs determined by their utility in alternative uses, etc., the one regarding the exclusivity of the function of money as a means of exchange is rather poor. Refuting the objective theory of value gave him partial justification. Value is and remains subjective. This is correct! Here, breaking away with the classicists meant revolution. However, his refuting of both the objective theory and the theory of money as a measure of value was a rather tenuous doctrinal gesture. Not even his allergy to Marx justifies it. The consequences of such a gesture would not have been so great if, from the position he held, he had not set the tone, the note and the method; he would not have called for a "closing of the ranks", for a subordination to the system. And it is known, the system freezes and kills.

1.1.2 Mises Versus Keynes at the Crossroad of Money That Does Not Measure Anything

Had they been convinced that the "leader" had solved the problem, the disciples would not have insisted. However, they did, and that is puzzling. The most persistent was Mises. The prominent theorist did not find anything tautological in the master's logic. He stops obsessively on the subject, although he would have had enough reasons to think, to the benefit of his own demonstrations about the unstoppable roll of credit out of nothing that something does not add up in Menger's design. In *The Theory of Money and Credit* (Mises, 2013) the sorcerer's apprentice does not even give the impression that he should be dealing with measuring value. In the spirit of the school to which he belongs, he does not take value for granted, but considers it as "the result of evolution"; the value is subjective, it is estimated, it "is not measured, but graded" (Mises, 2013, p. 47). Moreover, he is convinced that "[i]n any case the usage [of money as a measure of prices] certainly cannot be called correct" (Mises, 2013, p. 49). In his attempt to convince, he relies on a metaphor and argues that granting money the function of being a measure of prices is the same as "the determination of latitude and longitude as a 'function' of the stars" (Mises, 2013, p. 49). Considering the problem neither important nor scientific, he leaves it to the jurists! They should deal with "accidental" issues, not with "essential parts of the concept of money", Mises argues. The real problem is money conceived as a means of exchange. Here, he is interested in "the price of money", "objective exchange-value" and "the original starting-point of the value of money [that] was nothing but the result of subjective valuations" (Mises, 2013, p. 121). He consolidates these notions within the perimeter of monetary quantitativism, as he definitely strikes a chord with its principles. In line with this philosophy, he expresses his belief that "an increase in the quantity of money results in no increase of the stock of consumption goods at people's disposal" (Mises, 2013, p. 208). In other words, Mises breaks the connection between money and the volume of goods. If he is interested in something, that something is the stability of the monetary unit, of the standard of measurement. This is what matters, not the standard value of money. And this is because, he writes confidently, "endeavours to increase or decrease the objective exchange-value of money prove impracticable" (Mises, 2013, p. 236).

Not only the obsessive comeback to the issue of the functions of money induces the idea that here the foundation is not very sound; that clear judgements stumble into logical inadequacies. For example, Mises argues at one point that "*The purchasing power of money is the same everywhere; only the commodities offered are not the same*" (Mises, 2013, p. 176). How would Smith have reacted hearing such a thing?! Perhaps the function of money as a means of exchange manifests in the same way everywhere; the function is the same everywhere, not the purchasing power!

Such examples could go on to reinforce the same conclusion. To show us how a great researcher can go on two paths at the same time, without possibly giving up any of them to find a unique route. In other words, the text we quoted shows that Mises—like many others—could not completely dispel the Ricardian idea of money as a measure of value. After all, accepting this idea would not have entangled his analysis or conclusions. The acceptance of the primary function of measure of value does not imply the refusal to see in value a subjective measure. Even if we assign the market the task of assessing by means of "grading" values, in the end we still have to measure them. And if we do not use money to do so, what else should we use? It is difficult to understand how someone who tried to prove the unviability of socialism by the impossibility of resorting to monetary calculation, accepted the idea of breaking the value of goods from the primary function of money. Insisting upon the uniqueness of the function of a means of exchange, even if he did not aim to do so, he set the grounds for a phenomenon he feared and to which he sought, elsewhere, barriers: the unlimited expansion of credit. Since, essentially and strictly causally, what behavioural algebra does money as a measure of value compel to? What does it actually mean? It means: more goods → more money; fewer goods → less money. Money as a mere means of circulation means "injection" under any circumstances. The economy can contract but you can send more money! It is true, not to measure but to "exchange". In fact, in order to "save", as it turned out, those who had grown so big in times of crisis that it cannot be allowed to go bankrupt.

The clear and declared consciousness of this last reported phenomenon was embodied by J.M. Keynes. The author of the *General Theory* did not build a special argument dedicated to the functions of money. However, he did tackle the issue indirectly, when addressing the delicate problem of the interest rate. Indirectly, Keynes argues that "[i]f by money we mean the *standard of value*, it is clear that it is not necessarily the money-rate of interest which makes the trouble" (Keynes, 2018, p. 201; our emphasis). The problem is—we learn in the context—that *as a standard of value*, the elasticity of money substitution is null; we cannot substitute other factors of production with them. This is the problem! We learn about it, explained in the following text, even if the author thinks about gold as a standard of value: "It is interesting to notice that the characteristic which has been traditionally supposed to render gold especially suitable for use as the standard of value, namely, its inelasticity of supply, turns out to be precisely the characteristic which is at the bottom of the trouble" (Keynes, 2018, p. 207). So the "standard of value" is inconvenient. It cannot be lengthened, widened, shortened, adjusted, etc., as desired. And then Keynes is free to think that gold has become a barbaric relic; to dream that "if money could be grown like a crop or manufactured like a motor-car, depressions would be avoided or mitigated" (Keynes, 2018, p. 202); and, moreover, to imagine that "money is a bottomless sink for purchasing power, when the demand for it increases" (Keynes,

2018, p. 203). With no restrictions, Keynes urges us! From a "bottomless sink" we can feed the demand for money if the need requires it. And the great need is claimed in the name of unemployment. He could only get rid of it with a lot of money. "[M]en, [he argues], cannot be employed when the object of desire (i.e. money) is something which cannot be produced and the demand for which cannot be readily choked off. There is no remedy [pay attention to what Keynes tells us] but to persuade the public that green cheese is practically the same thing and to have a green cheese factory (i.e. a central bank) under public control" (Keynes, 2018, pp. 206–207). And the green cheese factory, the printing press was turned on, produced green dollars and functioned as a bottomless sink. It solved the problem in the short term. But in the long run, it sent us into a world of uncertain landmarks; in a world of shifting sands where playing with money, set apart from the value of goods, turns into a safe subscription to recurrent crises; to a lack of resilience!

Keynes was honest in motivating his pirouette. He was not "tied" to a school. He created it himself. Unprecedented circumstances were calling for radical action in the monetary policy. Everything possible had to be done to curb unemployment. Despite the tangled way in which he carried out his analysis, the message was clear, and he recommended it to the official policy. It is well known that what started out as a short-term solution turned into a new and tempting matrix: the Keynesian paradigm. Through it, and invoking the name of the mentor, the green cheese factory becomes an irresistible temptation during decisive moments. Even if Keynes did nothing more than to coagulate ideas that were floating in the ideational environment, the excess monetary issue, broken by the dynamics of the value of goods and manipulated by the interest rate is associated, paternally, with his name.

If in Keynes's case the ease with which he proposed the solution and gained his glory does not seem to contradict any precious scrupulousness, the situation is different for those who—playing on the same field—did not spare his alleged lack of rigor. We are again thinking about Mises. He was worried about the excessive issuance of money, regardless of its form, and especially the circulation credit, without support in voluntary savings. However, for the sake of the marginalist revolution, he argued, over and over again, that "Money is the thing which serves as the generally accepted and commonly used medium of exchange. This is its only function" (Mises, 1998, p. 398). Convinced that the *purpose* of money resided exclusively in its purchasing power, Mises believes that it is at the intersection of the supply and demand of money; just as all other goods or services are. This is his famous "monetary equation" whose analysis induces the main effects on interest rates and price dynamics. Prices that have nothing to do with any "measure of value" since "Money is neither an abstract *numéraire* nor a standard of value or prices" (Mises, 1998, p. 414). Based on such beliefs, Mises strives to reveal opportunities to limit monetary expansion derived precisely from the internal logic of the market functioning, and especially of banks. In a special subchapter entitled *The Limitation on the Issuance of Fiduciary Media*, obsessed with the danger of credit expansion, he optimistically—yet dubitatively—argues using a dozen of "if" that:

- Out of caution, the banks should be aware that "[i]t must not increase the amount of fiduciary media at such a rate and with such speed that the clients get the conviction that the rise in prices will continue endlessly at an accelerated pace. For, [he goes on to argue], (…) the public (…) will reduce their cash holdings, flee into "real" values, and bring about the crack-up boom" (Mises, 1998, p. 433).
- "The expanding bank must redeem its banknotes and pay out its deposits. Its reserve (…) dwindles. The instant approaches in which the bank will - after the exhaustion of its money reserve - no longer be in a position to redeem the money substitutes still current" (Mises, 1998, p. 434).
- "A bank can never issue more money-substitutes than its clients can keep in their cash holdings" (Mises, 1998, p. 435).
- "No bank issuing fiduciary media and granting circulation credit can fulfill the obligations (…) if all clients are losing confidence and want to have their banknotes redeemed and their deposits paid back" (Mises, 1998, p. 436).
- "If the governments had never interfered for the benefit of special banks, (…) no bank problem would have come into being. The limits which are drawn to credit expansion would have worked effectively. (…) Those banks which would not have observed these indispensable rules would have gone bankrupt, and the public, warned through damage, would have become doubly suspicious and reserved" (Mises, 1998, p. 438). If…!!
- "The establishment of free banking was never seriously considered precisely because it would have been too efficient in restricting credit expansion" (Mises, 1998, p. 438). If only it had been …!!!

- "[E]very restriction imposed upon the issuance of fiduciary media depends upon the government's and the parliament's good intentions" (Mises, 1998, p. 440). It depends ...!
- Etc.

As it can be seen, we have not exhausted the list of much-desired rule-like limitations, which, in the circumstances of an unrestricted banking competition, would hinder the lust for excessive issuance of fiduciary media. Mises knew that "Free banking is the only method available for the prevention of the dangers inherent in credit expansion" (Mises, 1998, p. 440), but he also knew that such a thing was not possible. He knew and he wrote that "Recourse to the printing press and to the obsequiousness of bank managers, willing to oblige the authorities regulating their conduct of affairs, is the foremost means of governments eager to spend money for purposes for which the taxpayers are not ready to pay higher taxes" (Mises, 1998, p. 440). What Mises also knew, and convincingly proved, is that the economic boom generated by credit expansion is, a priori, a bubble. Even if his explanation points to a cause that translates into "malinvestment" misdirected by entrepreneurs, this does not change things. Because misdirection is the result of deception induced by an interest rate manipulated by the unhealthy inflation of credit. In his own words, Mises argues that "What is needed for a sound expansion of production is additional capital goods, not money or fiduciary media. The boom is built on the sands of banknotes and deposits. It must collapse" (Mises, 1998, p. 559).

And then, if this is the end of the sandcastle built on issuance of shaky foundation, with the sole support of the marriage between the servility of bankers and the endemic populism of politicians, why feed on air? Why, if that is the *status quo*, should we consider Professor Irving Fisher's proposal that banks be required to maintain a 100% monetary reserve in relation to the amount of monetary substitutes to be "illusory"? Would this path be more dangerous than the free manoeuvring of the entire monetary and economic policy through the only but convenient and efficient means of which is the interest rate?

In other words, Mises gives us a wonderful lesson about the danger triggered by an expansionist boom. He sees all the limitations, but he knows that they are not functional. Moreover, he is suspicious about the one that seems to stand real chances; he deems it illusory. After all, he is consistent with himself. Throwing the "standard of value" to the dustbin of history and considering that the "only" function of money is that of means of exchange, he could not provide another analysis. An analysis from which we learn that money, instead of measuring, devours; no longer a mere unit of measurement, it becomes a manoeuvring instrument; in any direction. Keynes came to the same result, with fewer theoretical digressions.

1.2 Sustainability of Anti-crisis Therapy: Keynes Versus Hayek

1.2.1 Keynes: A Revelation and an Unclogged Road for the Problem of Sustainable Business Cycle

To be a product with chances to succeed in dealing with the problems of the 1930s, Keynes manifested as a synthesis. He remained a marginalist in as far as methodology is concerned, but a classic in the great judgements of economic analysis. He did not subordinate the real economy to the nominal one, nor did he make economic growth an end in itself. On the contrary, he was interested in the social problems he wrote about, and turned to the state to solve them; he did so by acting as a notorious socialist, although in his intimate structure he remains the student attached to Marshall's free market philosophy. Above all, he did not justify any means to his end—that of curbing unemployment and ensuring economic dynamics. His doctrinal scrupulousness did not stumble upon either the resort to the state, or to his giving up the barbaric relic—the gold standard. He was dominated by the vision of the whole, of the economic body as a system. If we want to find out how subsystems, parts of an ensemble, find their place and function within the whole system, Keynes is a qualified teacher. He uses all the variables belonging to the real and nominal economy. The "psychological incentives" complete his vision. And, above all, it is worth analysing his work in order to notice that there is—in the world of great economists—an unclogged end of the road in terms of the strong role of the natural rate, especially that of interest. From this point of view, it is worth finding out just how much there is to gain for the analysis of sustainability and resilience, via Keynes, correlated with Wicksell's work.

The real economy gives impetus to start a business cycle. The entrepreneur—interested in making an investment—is Keynes's starting point. He cannot do it by himself. An economy—full of competitors—plus a bank willing to lend are absolutely necessary. Keynes no longer raises the issue of the need for credit; he considers it normal. The entrepreneur asks for money and the bank offers it. However, not under any circumstances. "By the scale and the terms on which it is prepared to grant loans, the banking system is in a position, under a regime of representative money, to determine—broadly speaking—the rate of investment by the business world" (Keynes, 2013a, p. 138). A phrase worthy to be quoted; it would also deserve to be used as a slogan by any Central Bank! At the same time, the investment rate must be correlated with the saving rate. This is also the bank's mission. An investment is attractive to the entrepreneur only if the rate for borrowing, which is a cost to him, allows him to earn. However, a low, convenient rate for borrowing does not encourage the necessary savings. Gradually, Keynes realizes he needs Wicksell. He acknowledges him, in an appreciatory manner, and assigns to the bank the duty to determine the natural rate of interest. Keynes relies on three meanings of the natural rate of interest, just as Wicksell (1962) did. In addition, he explains and operates with this concept in a broader analysis, with interrelated—and most definitely inspiring—causal relationships.

A first meaning of the natural rate is that of the average rate of estimated profit. Its amplitude as such, a second meaning, is a result. "[T]he natural rate of interest is the rate at which saving and the value of investment are exactly balanced, so that the price level of output as a whole exactly corresponds to the money rate of the efficiency earnings of the factors of production" (Keynes, 2013a, p. 139).

The quote is too "full" not to be discussed. The desired profit is not a utopian issue. The extrapolation over time of an already established trend solves the problem. The evolution of prices, in relation to which entrepreneurs make their estimations, is also an indication for determining the natural rate. It is important that the bank, not just the entrepreneurs, be concerned with this issue. Investments must cover, through use, the savings. If this happens, it is because competition on the road of equalizing the factors of production leads to their natural prices and to levelled yield. No one receives more or less than the market "estimates" it is just, natural. That is what Smith and Ricardo argued in other words! In addition, if investment absorbs savings, then employment is also guaranteed to the greatest extent possible. The rate at which the fullest possible employment rate is ensured is the third meaning of the natural rate on which Keynes—following in the footsteps of Wicksell—operates.

Keynes considers it a great merit of the Swedish economist to point out that the interest rate influences the price level not directly but indirectly "by its effect on the rate of investment, and that investment in this context means investment and not speculation" (Keynes, 2013a, p. 177). The mechanism is simple. Putting the market rate of interest below the natural rate triggers a business boom accompanied by rising prices, and vice versa. Keeping the market rate below the natural rate with a view to invest more than the accumulated savings and thus gain growth is possible only with monetary increase from the bank. The road to inflation is open but not uninterrupted. Keynes sees the possible reduction in the interest rate as "a slight, long-continued drag"; a decrease in it translates into "profit deflation" or "income deflation"; declining profit mean inclination towards declining investment and, "so long as investment lags behind (or runs ahead of) saving by a given amount, prices will continue to fall (or rise) without limit" (Keynes, 2013b, p. 183). In the long term, he does not rule out that the price level be governed by the money supply. But, invoking the Gibson paradox, Keynes concludes that "[i]f the market rate of interest moves in the same direction as the natural rate of interest but always lags behind it, then the movements of the price level will tend, even over longish periods, to be in the same direction as the movements of the rate of interest" (Keynes, 2013b, p. 184). An idea that was less exploited both by academics and by those who, practically, inspired by Keynes, support more the monetary injection rather than the manoeuvrability of the interest rate. This in the conditions in which, remaining a Wicksellian, Keynes puts great value on the rate of interest as the main tool for regulating the economic dynamics.

Keynes considers all the situations in which the market rate is compared to the expected profit rate, with all its adjacent triggering mechanisms. In relation to Wicksell, he brings novelties regarding the movement of the natural rate. The movement is necessary when an increase in the market rate of interest affects the investment—savings equilibrium. The equilibrium is preserved if the natural rate increases at the same time; this is possible, Keynes argues, by stimulating savings or delaying investment. Wicksell warned that the natural rate must "drag" the market rate. Keynes adds that this is the case, but on a certain delay. This is due to the fact that the banking world does not quickly detect changes in the natural rate. The equilibrium savings = investments suffers due to this time lag. "In other words, when savings are abundant or deficient in relation to the demand for them for investment at the pre-existing level of interest, the rate does not adjust itself to the new situation quick enough to maintain equilibrium between savings and investment" (Keynes, 2013b, p. 182).

At the same time, a causal relationship in the circuit says that the number of jobs depends on the emulative context that profit creates. From this point of view, Keynes's bank has a very important role. This can be accounted for by the fact that a change in the bank rate induces a change in the market rate of interest compared to the natural rate, which will have an effect on the level of prices and, therefore, of profits. And the change in the profit rate determines the change of jobs offered by entrepreneurs. In explaining the relationship between the bank rate and the saving process, Keynes does not leave out his marginalist origins. Based on a Böhm-Bawerk-like logic, he states that "a fall in the bank rate required to stimulate investment so as to balance increased saving must (...) precede the increase of saving by an interval dependent on the length of the productive process" (Keynes, 2013a, p. 187). Also within the framework of Austrian marginalism, Keynes believes, just like Mises, that not all entrepreneurs value the chance of winning in the same way. Reasoning differently on the relationship between the market rate and the natural rate of interest, bankruptcy can only be the expression of some people's short-sightedness in building a correct business plan!

The movement of money as such is no mystery to Keynes. The effective aggregate demand, with both its components, consumption and investment, caught his attention. And not for another reason but for that of stimulating these two components to obtain the highest degree of employment. His *General Theory* is well known today. The "underuse equilibrium" as a de facto "natural" state of the economy remains a phrase present in all serious analyses. The multiplication effect of investment upon income, provided they are supported by a saving policy, is a sentence worth using in any discourse on sustainability. Linking the volume of savings to employment is just as important. The driver effect that the branches of the economy in which capital goods are produced have, through employment, on the other sectors of the economy, can be included within the same paradigm of sustainability. We could add here, under the same status, the sentence through which the ingenious Keynes explains the different way in which the amount of money and the preference for liquidity, via the "liquidity trap", influences the rate of interest. Finally, if someone wants to know how and in which direction move the gear trains of the complicated clock called economy—real and nominal alike—in order to shape the projection of a development with the attributes of sustainability, this is, without any doubts, the reference text: "For whilst an increase in the quantity of money may be expected, cet. par., to reduce the rate of interest, this will not happen if the liquidity-preferences of the public are increasing more than the quantity of money; and whilst a decline in the rate of interest may be expected, cet. par., to increase the volume of investment, this will not happen if the schedule of the marginal efficiency of capital is falling more rapidly than the rate of interest; and whilst an increase in the volume of investment may be expected, cet. par., to increase employment, this may not happen if the propensity to consume is falling off. Finally, if employment increases, prices will rise in a degree partly governed by the shapes of the physical supply functions, and partly by the liability of the wage-unit to rise in terms of money. And when output has increased and prices have risen, the effect of this on liquidity-preference will be to increase the quantity of money necessary to maintain a given rate of interest" (Keynes, 2018, pp. 151–152). The quote above is instrumental in illustrating everything or almost everything about how an economy can work—well or poorly.

What else is there to add? We could also mention the fact that the social problem engaged Keynes for economic reasons. The arbitrary and inequitable distribution of wealth concerned him to the extent that income had to reach the consumers. Hence the anathema on the rentiers. And this is not out of some attachment to the proletarian cause. A correction of income inequalities could also come from an active fiscal policy. The budgetary balance did not limit his judgements either. It remained, just like the gold standard, a dogma of the classicists. For Keynes, the budget deficit could very well be a source of funding. How attractively unsustainable has this idea become in the meantime? And what an echo has aroused his idea to stimulate even the unproductive expenses if employment requires it—let us take the example of digging the Treasury!

As for inflation, as a weapon against unemployment, we believe that Keynes did not say what others put in his mouth or attributed to his writing. He was, indeed, mainly interested in employment. For the sake of it, he sacrificed almost everything. He was convinced by the mercantilists he paid homage to that the main way to create jobs was through investment. We have seen, along the way, that he was considering real investments, not speculative ones. He considered the control of the rate of interest to be the most important tool likely to create a favourable environment for real, sustainable job creation. But not anyway and not under any conditions; however, this is precisely what we tend to forget. Persistently referring to Wicksell, Keynes correlated the market rate of interest to the bank rate; he clearly defined them and placed them both under the umbrella of the natural rate. In doing so, he did not admit just any game. He supported the need to keep the market rate of interest below the rate of profit through additional money issuing in order to keep the investment drive alive. He did not pay attention to every decrease in the market rate of interest. The adverse reaction to saving was a good enough indicator. Therefore, he made his great judgements under the authority of the natural rate. He thus advocated that the interest rate follows the natural rate, which, not coincidentally, he defines it as the one that maintains the equality between savings and investments or that maintains the highest employment rates. We know well that he was agreeable to the temptation of increasing the amount of money. The "liquidity trap", that he rightly theorized, lowered his enthusiasm. He agreed to pay for an increase in employment rates with an increase in prices. However, once the problem was solved, a new injection no longer seemed necessary. He says it clearly, without a doubt: "the scale of investment is promoted by a low rate of interest, provided that we do not attempt to stimulate it in this way beyond the point which corresponds to full employment" (Keynes, 2018, p. 333). That's it, nothing more! No need for endless cash excess. Compared to the above, does inflation have anything intrinsically Keynesian in it? By no means, we would argue. However, the deliberate distorted interpretation of Keynes's texts may lead to such conclusions.

1.2.2 Hayek: When the Sustainability of the Business Cycle Rests on Shaky Premises

1.2.2.1 Searching for Sustainability in the Hayekian Analytical Labyrinth

Three books, *Monetary Theory and the Trade Cycle*, *Prices and Production*, and *Profits, Interest and Investment*, place Hayek on the path connecting the natural price of the money to the business cycle. The three books serve our purposes as they powerfully encompass all the great ideas that are likely to nurture the economic dynamics in a contemporary economy. The laudatory or critical appreciation of the way in which Hayek addressed the issue of the economic cycle, the chances of resilience, or non-resilience, gather all the great minds of the past century as well as of the present. He set the foundation of a solid construction that we will also rely on, looking for what deserves to be preserved as a sustainable foundation to build on, and which are the areas where "playing with money" must be avoided.

What exactly is Hayek interested in? He is interested in:

- The economic dynamics with chances for sustainability based on voluntary savings and the economic dynamics based on the "fatal conceit" of credit out of nothing, with the inevitable end in a crisis;
- The starting point of an "adventure" of the economic expansion in relation to the state of equilibrium;
- The importance of the rate of profit in the business cycles;
- The outline and content of the route covered by the relationship price-money-production structures from the beginning to the more or less happy end. Within this context, the turning point reached by balancing out the bank rate with the natural rate of interest is particularly noteworthy;
- The role of the central bank and the commercial banks in the process of driving, supporting or dynamiting the economic dynamics;
- The "mystery" of effective demand, the important role of the distribution of the quantitative surplus of money to the two main directions: consumption or capital formation;
- Sources of economic crises—the monetary and non-monetary ones;
- What is to be done? Before, during and after the crisis!

The brief work agenda is sabotaged by a couple of assumptions related to the methodology used by Hayek. By far, the idea that the function of a means of exchange summarizes the *raison d'être* of money dominates the analytical landscape. Doctrinal whims keep it away from money as a standard of value, although, indirectly or even subliminally, just as in the case of other Austrians, this function chases his analysis like a shadow. Then he refuses any type of "average"; the aggregate parameters irritate him. The serious enemy is the "general level of prices". He knocked it off the stage, "militarily": "theories which explain the Trade Cycle in terms of fluctuations in the general price-level must be rejected" (Hayek, 1933, p. 106). Menger's style can be easily recognized here! In *Prices and Production*,

Hayek mentions Wicksell, but not in order to build further on his theoretical foundation, but to find that, by a "curious irony of fate", he "has become famous, not for his real improvements on the old doctrine, but for the one point in his exposition in which he definitely erred: namely, for his attempt to establish a rigid connection between the rate of interest and the changes in the general price level" (Hayek, 1967, p. 23). It thus convinces us that in the least probed and most labyrinthic places of economic theory we cannot rely on him when looking for arguments for sustainability! Why? Because the "erred part" is related to the "general price level"; which, Hayek argues, does not exist! A "strong" demonstration is usually meant to clarify. However, in his case it is not clear. Here, in *Prices and Production* is where Hayek replaces the "natural rate" with the "equilibrium rate". By the way, for Wicksell, the natural rate also had this meaning. Then, with reference to the three hypostases in which the natural rate can be compared to the money rate, Hayek briefly explains what the Wicksellian "cumulative process" would have meant. He concludes that it was not Wicksell who said the right thing, but Malthus! It is not the "general price level" that is the hard pill to swallow, increasing or decreasing, but the level of prices for capital goods or consumer goods! However, aren't the latter also in the category of "general levels"? On the path of his own analysis, we notice that Hayek is no longer put out and makes use of averages, the average levels of capital prices or consumer goods! He even points out the merits of Mises, arguing that he "has improved the Wicksellian theory by an analysis of the different influences which a money rate of interest different from the equilibrium rate exercises on the prices of consumers' goods on the one hand, and the prices of producers' goods on the other" (Hayek, 1967, p. 25). Between the need to remain loyal to the School and the need to be rational, he is forced to appeal exactly to what he threw away in order to be convincing. He uses the natural rate of interest under the terms put forward by Ricardo and Wicksell in *Monetary Theory and the Trade Cycle* and *Profits, Interest and Investment*; in *Prices and Production* he calls it the "equilibrium rate". If more confusion was required in the notional system operating on the subject, he surely fulfilled his duty! Price stability, as well as the intrinsic value of money are not major objectives. However, the quantitative volume of the currency is indeed a major objective! He is concerned with the reserve fund, the old Smithian obsession. He also draws attention to the distinction between voluntary savings and involuntary, "forced" savings, which result from expanding the amount of money. He relies on

Böhm-Bawerk's "stages of production" and "roundabout way", as well as on Wieser's classification into specific and non-specific goods. It attributes a special meaning to used and unused resources, while the classification of investments into new, ongoing or complementary helps him appreciate the length of time between the implementation of the reserve fund and the turning point in running a cycle. He builds a trade cycle based on the dichotomy of industries into consumers' good and capital good industries and, moreover, on the differences within the latter group which he classifies it according to the more or less "capitalistic" type of machinery, and this depression and revival is not due to changes in the rate of interest.

Faithful to his creative processual character, with returns to some ideas that, in the meantime, had already received consecration, we start with an analysis of the main opinions in *Prices and Production*. Here, Hayek offers four famous Lectures: on dynamics, sustainability, and economic resilience. To point the direction he has in mind, he accompanies each lecture with a "motto", texts signed, in order by Cantillon, Malthus, Mises and Robertson.

The first lecture is a warning designed to show us what he agrees with or disagrees with, and point out the foundations of his analysis. Here we see that the sustainable Hayek is not always what we want: a statue to worship, no matter where you're standing. Dilemmatic and often at odds with his own hypotheses and working concepts, he does not send us to a solid ground where we can feel safe on our feet regardless of the winds that flutter our thoughts. He puts forward a systematic analysis of the evolution of monetary theory and he seeks landmarks. It is an opportunity to make a generous inventory of ideas, one more sustainable than the other. He is drawn to Cantillon for two great ideas: (a) the logical and complete explanation of the causal chain between the amount of money and prices; (b) the positive effect of the money surplus for those whose incomes increase first, in relation to the losing effect on those whose incomes increase later (Hayek, 1967). He takes from D. Hume a key idea in explaining the dynamics of the economic cycle, namely that the quantitative surplus of money has positive effects only in the interval between the "injection" and the increase in! (Hayek, 1967). The two authors mentioned are related to the standard theory that Hayek believes has three major errors: (1) money influences prices and production only by mutations to the general level of prices; (2) increasing prices tends to lead to an increase in production, and vice versa; (3) the monetary theory is reduced to a theory of the value of money.

Hayek dismantles the first hypothesis easily and convincingly: there is no tidal effect in prices—this has already been stated before. If all prices move in the same direction, simultaneously and equally, the effect is zero. There are "delays", "frictions" between the different types of price—depending on the goods prices refer to. If something is important and meaningful, that something refers to relative prices; they "determine the amount and the direction of production" (Hayek, 1967, p. 28). The "general level" and the "total production" cannot be reconciled with individual decisions. According to Hayek, relative prices referring to consumer goods or capital goods constitute a head start over the classicists. In order to keep the progressive trend, "monetary theory will not only reject the explanation in terms of a direct relation between money and the price level, but will even throw overboard the concept of a general price level and substitute for it investigations into the causes of the changes of relative prices and their effects on production" (Hayek, 1967, p. 29). A great prospect, but not even him contributes to imposing it. Relative prices apply to goods belonging to the same category. It is not at all difficult to see that when formulating his judgements, he does so in terms of the average prices of consumer goods, respectively, capital goods, even if he does not really use the term. The statement regarding the "interspatial" and the "intertemporal" equilibria of relative prices is, in theory, acceptable; but it does not spare him and does not protect him from the inertia with which he used the average parameters, despite a formal refusal.

He is labyrinthine and unwillingly nebulous in the position of dismantler of the third "error". The concept of value, in general, and in particular the value of money does not tell him anything. Or, better said, what it does tell him upsets him so much that he bluntly argues that this is not his problem, nor is it ours, but the question of whether or not "the state of equilibrium of the rates of intertemporal exchange is disturbed by monetary influences in favour of future or in favour of present goods" (Hayek, 1967, p. 30). Hayek ventures and believes that "[t]o discover the causes why certain needs, and the needs of certain persons, can be satisfied to a greater degree than others is *the ultimate object of economics*" (Hayek, 1967, our emphasis, pp. 30–31). His stubbornness in supporting this idea, refuting the theory of objective value and its "dangerous" resemblance to monetary theory makes him look tough. One would expect him to be credible to the end. The manner of argumentation and the unnecessary amount of energy spent on this subject, to which adds the nebulous proposal to fasten the object of economic science within an area of behavioural psychology and the hierarchy of need "softens" his sustainability. There's nothing left of his sustainability when he apodictically ordains that "the absolute amount of money in existence is of no consequence to the well-being of mankind" (Hayek, 1967, p. 31). Hayek wanted to express opposing views from his colleague from the London School of Economics. With this last sentence, fellowship with Keynes is sealed! Not so much in the way of understanding the role of currency as in the message sent for the normative environment.

With the second "error", Hayek also sends us into the blurry world of denying all assertions and all negations. He himself assimilates the Ricardian and Wicksellian idea of the suggestion that a future price increase determines entrepreneurs to invest. And if they do so, the result is an increase in production. What is wrong here?! Apparently, nothing!

Mises's containment to settle for just explaining how the effects of a quantitative surplus of money are distributed differently between different spheres of activity is considered to be a "not unimportant defect" (Hayek, 1967, p. 11). Such a harsh verdict attributed to his teacher seems to be accompanied by promises. Some will come; but most are illusory. Hayek will focus on this topic. He considers Cantillon's warning to be very important. It is thus important to know the "place of entry (or exit)" of the money surplus; it is even more important to know whether "the additional money comes first into the hands of traders and manufacturers or directly into the hands of salaried people employed by the State" (Hayek, 1967, p. 11). The economic cycle mechanism, redesigned by Hayek, brings enlightening pluses to the effects caused by inputs and outputs. As for the starting points, Hayek excelled himself in multiplying them and sending the beginnings of business cycles in areas belonging to a variable geometry that is hard to understand.

In search of inspiring sources, Hayek is willing to rely heavily on Henry Thornton's ideas. He quotes him, referring to his work *Paper Credit of Great Britain* in the footnote of page 13 of *Prices and Production*, to give the impression that he is seduced by the idea that "As soon, however, as the circulating medium *ceases to increase*, the extra profit is at an end". Another idea attributed to Thornton regarding "the influence of an expectation of a rise of prices on the money rate of interest, a theory which later on was to be re-discovered by A. Marshall and Irving Fisher"—is important, Hayek believes, but "does not concern us here" (Hayek, 1967, p. 14). It is not because, if it were, then Hayek would be forced to account for the logical inaccuracy of the second "error". Instead, he admits that the idea received "a more modern air" through Ricardo when he sent it into the analysis area regarding "rate of interest falling below *its natural level* in the interval between the issues of the Bank and their effects on prices" (Hayek, 1967, our emphasis, p. 14). Hayek will only briefly discuss this idea. He will talk about the level of equilibrium, and with this, hard to believe, in this work, the tribute to Ricardo ends here. An author who was the subject of the Wicksellian synthesis is sent out! Too bad, because the idea above, together with the one regarding the disappearance of extra-profit once the growth of the amount of money stops, are part of the strong ideational core of the business cycle theory.

He could not leave out another few names, right before setting out for his own analysis. Jeremy Bentham is one of them. Bentham's "Forced Frugality" spoke of a "future wealth" gained through the use by a government, either of a fund collected from taxes and fees, or through a creation of paper money. "Forced Frugality" was a far too generous concept to be fully exploited. Hayek does not do it. He is just following the widespread path of "future wealth" opened by forced saving. Within this context, he gives Walras credit for having pulled the term from oblivion and having sent it to Wicksell. The other path does not concern him here. Pascal Salin will also turn it into a consistent subject in *Tax Tyranny* (Salin, 2020). In fact, the same direction—which, through Laffer, becomes the doctrine of supply—is at war with this "incestuous" way of robbing the individual and then promising that thanks to the skilfulness of the state, the fruits of wealth will rise from robbery.

Malthus is another name that invites Hayek to reflection. Starting from the idea that "a change of the proportion between capital and revenue to the advantage of capital (...) would have the effect that in a short time, the produce of the country would be greatly augmented" (Hayek, 1967, p. 19) he will go on to build a whole wealth production mechanism in which it is not the volume of money but the distribution of funds between capital and income that carries the most weight. Besides Malthus, the classic Mill is not ignored either. He is mentioned three times. Once when in the *Principles*, he leaves room for an "emasculated form" of the interest natural rate (usual, calls it here Hayek) (Hayek, 1967, p. 17); the second time, as he attributes to bankers the ability to convert income into capital and, in the name of monetary depreciation, forced accumulation to stimulate the economy; and finally, the third time, he quotes from an angle that is convenient to his analysis, where Mill says that "there cannot, in short, be intrinsically a more insignificant thing, in the economy of society, than money" (Hayek, 1967, p. 126).

It is not easy to guess the key in which Hayek drew his theoretical offer in the area of the economic cycle and monetary theory. The effort to find it is subordinated to the goal of foreshadowing an economic policy that avoids turbulence, both in the monetary area and in the real economy. For such a goal, we need supportive arguments and we are looking for them, naturally, in the works of the great scholars. And what do we find in the writings of Hayek?

As a general frame, through the analyses carried out in the second and third of his famous lectures of *Prices and Production* and *Monetary Theory and the Trade Cycle*, Hayek intends to continue the work of Malthus and Mises. Nevertheless, he does not do it in just any way, but by relying on Böhm-Bawerk. Ensuring an equilibrium between the production of consumer goods and that of capital goods seems to him to be essential for a sustainable dynamic. The concrete mechanism by which a surplus of money reaches both directions—consumption or investment—and whether or not, by means of the rate of interest, it manages to fulfil the conditions of the above-mentioned equilibrium, also seems to Hayek an attractive analysis attempt.

A starting point is the intention to answer the following Malthusian interrogation "The question of how far, and in what manner, an increase of currency tends to increase capital [provided that, he adds,] [i]t is not the quantity of the circulating medium which produces the effects here described, but the different distribution of it (…) on every fresh issue of notes (…) a larger proportion falls into the hands of those who consume and produce, and a smaller proportion into the hands of those who only consume"(Hayek, 1967, p. 32, extract from the motto prefacing the second Lecture—quote from Malthus). Standing with Malthus and following in the footsteps of Böhm-Bawerk, Hayek aims to prove that the capital/consumption ratio is a function of each production stage. And, if this is true, what is it that sets things in motion? What is it that makes the entrepreneur want to multiply the stages of production and how do consumers interfere with this approach?! To find the underlying cause of it, Hayek is at war with many and adopts a variety of hypotheses, some more "heroic" than others. We believe that there was no need for many digressions to get to the core. And we also believe that it is precisely the excess of contradictory hypotheses and the untruths pronounced as legalities (see, for example, the assertion that "the main causes of variations of industrial output are to be found in changes of the willingness of individuals to expand effort (...) it is a highly artificial assumption to which I would only be willing to resort when all other explanations had failed"!!!, Hayek, 1967, p. 33) that sabotaged the core of the analysis.

The six graphs in Lecture II and the diagram in Lecture III direct us straight to the core. A place where the "mathematician" Hayek outdoes himself, with four variables on two axes, explains how the impossible can become credible. That is, he tells us, nonchalantly, that "within practical limits we may increase the output of consumers' goods from a given quantity of original means of production indefinitely, provided we are willing to wait long enough for the product" (Hayek, 1967, p. 38). We also learn from him that waiting is not enough. Then, Hayek proceeds—in a direct and Böhm-Bawerk-like style—to the analysis of the way in which the transition from one method of production to another involves lengthening the period of production, as well as changes in its structure, with effects on the ratio between capital goods and consumer goods. The graphs, just like Lecture II as a whole, are meant to prepare the interested reader to understand the analysis carried out in Lecture III. What is the holdup here? We have learned that using sage graphs, Hayek tells us that individuals do not consume intermediate products; that the products as such represent nothing if they do not lead to consumer goods! "Father" Malthus would have said: ask and you shall receive! Here are the words Hayek uses to say this: "... the proportion of money spent for consumers' goods and money spent for intermediate products is equal to the proportion between the total demand for consumers' goods and the total demand for the intermediate products" (Hayek, 1967, p. 46).

How to move from one method that uses less capital to one that uses more capital seems to be the essence of Hayek's response to the Malthusian interrogation. The key sentence is: "The continuance of the existing degree of capitalistic organisation depends (...) on the prices paid and obtained for the product of each stage of production and these prices are, therefore, a very real and important factor in determining the direction of production" (Hayek, 1967, p. 49). It is, Hayek rightfully believes, up to every entrepreneur if "he re-distributes his total money receipts in the same proportions as before" (Hayek, 1967, p. 49). The "expected level of profits" will influence his decision: increase production by saving more, or consume like before, keeping the scale of production and the production method. Only in the "dialogue" with Mises does Hayek respond clearly to this challenge. Here, he does so "suffocated" by the "average period" of production, the waiting time, the "average number" of transactions, stages, etc., that is, exactly what he declaratively does not like.

We would have enjoyed, here, another—or better said—yet another type of analysis! If at each production stage, the capital/income ratio is a function of each entrepreneur and each makes decisions based on the "expected profit", is it possible that he spoke of a range of natural rates in relation to the order of the stages of production methods? Definitely! A kind of law of diminishing returns occurring with the extension of the production period is considered by Hayek in another context. However, this was also a context in which he could have questioned himself on how the different natural rates would have led to different decisions on whether to continue production or not. Alternatively, a possible conclusion could have been the following: if you do not have competitive markets at every stage, you may hit a wall! Faithful to Böhm-Bawerk, instead of such reflections, Hayek concludes that "As the average time interval between the application of the original means of production and the completion of the consumers' goods increases, production becomes more capitalistic, and *vice versa*" (Hayek, 1967, p. 42). When that "vice versa" happens, crisis awaits, he believes. Hayek's conclusions seem credible—up to a point. Until we remember the "super-heroic" starting hypothesis: we do not take into account technical knowledge (see Hayek, 1967, p. 93, footnote). It is only on such a premise that Hayek's graphs make sense. With technical progress, his graphs just "blow up"! And so does his conclusions.

To sound more convincing, Hayek changes his plans. He leaves out the graphs and uses the classification of goods into "specific" and "non-specific" initiated by Wieser. It retains only formally the meaning of its predecessor's classification. On the same assumption of the neutrality of technical knowledge, Hayek's goods move freely between production stages as dictated by the profit-led interests of the entrepreneurs. Regardless of their nature, these goods move towards "earlier stages". "[T]emporary differences between the prices" have their own trajectory. "Price margins" are subject to a legitimacy of diminishing returns. They exist, otherwise the reason for permanent reinvestment would disappear, but the tendency is to reduce their amplitude gradually. In addition, a turning point—a sort of "returning to origins"—emerges on this route, based on the hypothesis that consumers (with increased wages at the beginning of the boom) decide to save and invest part of their income. The effect on the prices of capital and consumer goods determined by the change in the demand–supply ratio for the mentioned goods is well known. Although he claims to do so, Hayek does not make considerable additions compared to Wicksell and Mises. However, he does notice that in this movement of goods and services determined by the variation of relative prices from the various stages of production to the equalization of profits at all stages, there are places worth paying more thought to. Such a place would refer to the natural rate. He believes that the (natural) equilibrium rate, which involves capital absorption of all voluntary economies and stability of the "general level of prices", as it appears in the Wickselian scheme, would be characteristic not of an equilibrium state but of a stationary one. Or, we know that not the stationary state but the "transition", the movement—including of prices—pertains to what is natural. And movement, dynamics has its limitations. Limitations that Hayek is willing, or not, to see.

At times, Hayek believes, with undisguised enthusiasm, that the production period can be extended indefinitely. If the producer "dies" for the sake of the consumers and wants to cater to their needs with more and more durable goods, even if the advantageous price margins decrease, we allow ourselves to agree with him. Just as we can agree with him if we imagine the unimaginable: that there are no stages we can overlook; technical specificity and technical progress are missing from the game! But on what ground would we find ourselves speculating in this manner? On the one of sustainability, certainly not!

The period during which the monetary surplus of an "injection" becomes income also depends on the size of the "reserve fund", which is, actually, correctly grasped. It is important, after starting a business, to have your own money to pay salaries and avoid paying them off the loan. However, after the first series of finished products is sold, Say's Law comes into effect. The supply created brings about the potential demand. One no longer pay salaries from the reserve fund. The balance of prices for capital goods in relation to that of consumer goods could benefit from the logic of the well-known classic. Neither Hayek nor his contemporaries find it necessary to claim it—as a name or as an idea. A sequence of events following the pattern: *first* increases in the price of capital goods and *then* in consumer goods would have lost some of its clumsiness, and not just methodological.

Hayek set out with the intention to further substantiate the idea, relying on Cantillon and building on the works of Malthus and Mises, that it is very important to know where the money comes in and out of the economy. It is one thing when the money goes first to those who produce, and another is when it first reaches the civil servants. A mechanism that may function based on such a premise seems to be a seductive logic. Nevertheless, we believe that it is not only unattractive to logic but also dangerous. There is no reality in which "fresh money" first reaches at …! Would someone just throw it, just like that, and then the strongest are the first to eat?! And then, what reason would explain why the money goes to civil servants first? To civil servants in their capacity as consumers or as voters? The representatives of the New Institutionalist School analysed the "make-or-buy" hypothesis—rather "brave", by the way. At the same time, as we have already shown, Mises demonstrates that you first have to be a "producer"; earn money, and then you can afford to be a consumer. Otherwise, if you receive the money first, you become an office clerk. This is the hypothesis that Hayek is playing with. Can something like this happen during a cycle? Electoral reasons can send money to voters interested in this deal. They will be self-loyal and self-clerking. The Austrians' aversion to the bureaucratization of the economy is well known. However, just toying, even if only for the sake of analysis, with the idea that something like this could happen, that from a large pile, money first reaches the civil servants, is risky. Any government is happy to quote any work that, academically, "logically" argues in favour of distributing money first to the civil servants. Rationally thinking, an economic cycle cannot set its start on such a premise. The money must reach those who ask for it for business and who, starting a business, will create jobs in the real economy; with potential chances to sustainability, according to a simple mechanism also borrowed from Say. That should have been Hayek's cutting position. Not the doubtful possibility that the beginning belongs to civil servants.

Equally, the fact that the starting point in the business cycle is a "slippery slope", sometimes for entrepreneurs, sometimes for the central bank, is not one of the strong points in Hayek's theory. For those in the classical school, it was clear that the wand belonged to the entrepreneur. The entrepreneur is not absent from the Hayekian landscape but is not always a leading player. The power of the bank to create credit according to its own interests may overshadow its status. And it does so both at the beginning of the road and along the way. Its "ward" seems to be fatal.

Why? Because the economy is not satisfied with the state of equilibrium; it wants transition, and it wants more and more! Moreover, unfortunately, it cannot fulfil this desire on its own. Sustainable development appears in Hayek's eyes as one driven by voluntary savings. Correct, but it is not enough; neither in relation to desires nor to people's propensity to save. "Had saving preceded the change to methods of production of longer duration" everything would have been fine and sustainable (Hayek, 1967, p. 88)! However, he sighs, "this necessity will be resisted" (Hayek, 1967, p. 88). The Bawerkian preference for the present induces the "immanent" endemic need of loans. It is not the voluntary savings but the forced savings that define "normality". Increasing the volume of money is both fatal and natural! Hayek believes that Wicksell erred in associating the stationary state, that of equilibrium with normality. The Wicksellian situation in which the loan interest rate equals the natural rate, the savings are completely absorbed by investments and the currency becomes neutral in relation to the general price of goods, seems to him contradictory and, why not, exceptional; exceptional because the economy is, he believes, expanding. Moreover, he also believes that "[*t*]*he rate of interest at which, in an expanding economy, the amount of new money entering circulation is just sufficient to keep the price-level stable, is always lower than the rate which would keep the amount of available loan-capital equal to the amount simultaneously saved by the public*: and thus, despite the stability of the price-level, it makes possible a development leading away from the equilibrium position" (Hayek, 1933, p. 114).

Despite his complicated wording, Hayek argues that a general price level remains stable not at the parity of the lending rate with the natural (equilibrium) rate. A "general price level", as we have already noted, does not exist for Hayek. What matters is the volume of money supply and the level of prices for capital or production goods. Tampering with volume leads to cyclical economic and price fluctuations. And volume is tampered with because voluntary savings are never enough. The tampering is no mystery. In other words, Hayek says exactly what those who prefaced him said. He emphasizes that cyclicity has an endemic, endogenous nature. It is not like outsiders come to move us cyclically; we do it ourselves! We do not have enough savings and we borrow money. The elasticity of the money supply allows us. The drag, by manipulating the rate of interest, lasts as long as the bank allows it. It usually does not hold. Fluctuations are therefore inevitable. And, basically, no one is to blame for what is happening. It would be "nonsensical (...) to formulate the question of the causation of cyclical fluctuations in terms of 'guilt'" (Hayek, 1933, p. 189). The cause belongs to us. Conclusion: "So long as we make use of bank credit as a means of furthering economic development we shall have to put up with the resulting trade cycles. They are, in a sense, the price we pay for a speed of development exceeding that which people would voluntarily make possible through their savings, and which therefore has to be extorted from them" (Hayek, 1933, pp. 189–190). And we always resort to bank loans. Not because the bank invites us, but because we want to grow faster and more than our voluntary savings allow it. Otherwise, banks would only explain their existence as safe deposit or as payment intermediaries. Could such an image complete the picture of the catastrophic supporters of zero growth or decline? Can Hayek inspire them when they spend so much on the idea of the need for a constant volume of money? Basically, Yes!

The stubborn refusal of the function of money as a measure of value leads Hayek, indeed, to an area where he can only discuss with the proponents of zero growth. He disavows the "fanatics" of the standard of value. Not Ricardo or Marx—emblematic examples on the subject—but he chooses Cassel and Pigou as losers of this temptation, and criticizes by deliberately misinterpreting them, in a tirade where the general level of prices "kills" him again. He admits a permanent variation of the amount of money in relation to a variation of production only at world level. An embarrassing exception caught in an equally lame logic. At country level, only mutations in the "coefficients of monetary transactions" (clustering, dislocations, etc.) or at the level of the speed of money circulation can place production in relation to the amount of money. Otherwise nothing! This contradicts the idea—dear to his heart—that "money will always exert a determining influence on the course of economic events" (Hayek, 1967, p. 126). That is, the currency is active, not as Mill thought. However, how can it be active if it bustles with the variation of production only at the world level? According to Hayek's logic, at the country level, we should theoretically have prices without money. Practically, is this impossible because we have to sign long-term contracts expressed in this "general medium of exchange". Or, the contracts in question are expressed in prices which reflect the value of the goods. He thrives in the example—so well analysed—of the expansion of private credit and especially of credit money, on the uncontrollable discount; money "grows" here beyond variations in production. However, he falls for it, for the sake of the idea. Credit money successfully replaces real money as a means of exchange. Only "sooner or later", this money must be accounted for; it will have to find a counterpart: express values or mean nothing. In full crash, this is what happens, it no longer expresses anything, and it means nothing. Moreover, in the midst of the crisis, individuals are sending their money to safe havens, precisely because they are the ones to preserve it; their prices remain standing, sustainable, beyond what Hayek believes is going on in the world of commodities with values that are not expressed in money, but directly in prices. A world where, according to him, we can leave out the money, the prices remain!

The world of money movement belongs to the world of the movement of goods; not the other way around. In this movement, in order to reach a "natural result", not only prices must be brought to their natural level, but also quantities. Money is used to obtain as many goods as possible; not inversely. Within this process, money plays an active role. In relation to the founders' beliefs that goods are produced by goods with the help of money and not that money is produced by money beyond the world of goods, the idea of "neutral money" is, if not illogical, at least stupid. "Specializing" in money by forgetting or eluding the world it has to serve is just as stupid.

Just like in the case of the other Austrians loyal to Menger, who reject completely the value of money, Hayek places himself in the position of a prisoner of his own contradictions. He criticizes the "eminent" A. Smith, but he wakes up in his court when, on page 45, with an explanation faithful to the criticized, he writes that the "output of consumers' goods is necessarily equal to the total income from the factors of production used, and is exchanged for this income" (Hayek, 1967, p. 45). We can recognize here Smith's dogma, the value of the sum of the three incomes, as Marx would remark, just as critically. There was no need for the contagion with Marx to explain what he was thinking. The reference to Marshall would have been enough, and then he would not have formally rejected the costs placing them into cumbersome explanations. Or, he would have told us, in the context of his theory, something personal about the danger of falling prices against the background of rising labour productivity. Because Marshall, which he cites in the footnote on page 106, spoke of a decrease in costs as productivity increased. Reduced costs mean extra chances for competitiveness. Every entrepreneur is interested in such a thing. That these costs add up to prices (and not the other way around!) is an idea dear to Hayek as well. But when any entrepreneur seeks to cover his expenses (costs) from selling his products at a remunerative price, the dialectics of a reverse causal relationship is less interesting; or not at all.

1.2.2.2 Are Our Expectations Too High?

In one sentence, we would have liked to answer the title question as follows: no, no expectation from Hayek is too high. However, our author does not fully put us in this satisfactory position.

Let us start, however, positively. The insistence with which Hayek tries to convince that the source of sustainable development lies in voluntary savings is to be appreciated. It is the "capitalistic" way, as he calls it, by which money is sent for capitalization; because, logically, one reduces the part that could be intended for consumption in order to invest it; invest it to increase consumption in the future. The extension of the production period is ensured.

However, voluntary savings are not enough for the capitalist "lust" for development. Forced savings are needed. Forced saving is not happiness but a necessary evil. Investments above the level of voluntary saving are needed all the time; for objective but also subjective reasons. If this is the case, their mechanism must be mastered and helped to lead "to a certain result" (Hayek, 1967, p. 151). Hayek views this result in terms of a "constantly increasing rate" of the capital. The way to reach such a desirable goal is an injection of new money. It comes from the bank in the form of a loan, but someone calls for it. We have seen that entrepreneurs and civil servants participate in this game. And we have seen that it is important who gets the money "first"! In this context, the fact that Hayek was concerned with the direction taken by income, once obtained, towards consumption or saving, is a commendable thing. No novelty here. Compared to those who preceded him, he set out to find an optimization scheme for this proportion. But he remained in the intention phase. This is because we are dealing with an impossible mission, he states with disappointment!

Unsustainable for sustainability is the mechanism itself, fuelled by forced savings. Hayek proves that price structures influence production structures. Despite too "brave" assumptions, he credibly argues that the ratio between the prices for capital goods and consumers' goods is very important for maintaining the economic dynamics. However, prices mean money and it is desirable, he believes, with a few notable exceptions, that the volume of money in circulation remains constant. An impossible requirement as long as we accept that the monetary quantitative surplus has an "endemic", endogenous motivation. It would be normal and natural for those who ask for money to do it rationally, moderately. A "legitimate demand"—which stirred smiles, as we find out also thanks to him—should have been linked to an equally legitimate bank offer. The natural rate of interest mechanism, so skilfully handled by Ricardo and Wicksell, does not inspire confidence. Although his analytical course does not leave out the entrepreneurs' concern to make calculations based on the evolution of prices or expected profit, when he has to put his finger on it and resort to this remarkable lighthouse in the industrial orientation, Hayek abandons. Abandonment that equals not only failure, but also huge steps back from a constant and consistent classical and neoclassical effort to find a steady line by calling the competitive market to determine what is normal, natural and acceptable for each to receive. The natural price of money, called "natural rate of interest" has precisely this role. To satisfy those who ask for money as well those who offer it. Moreover, pinning it at the level of the expected average rate of profit is a principle with the vocation of synthesis; one of temperance and balance. The fact that he correctly established that it was not equilibrium but the tendency towards it that represented "normality" in the economy did not entitle him to dilute the outstanding role of the natural rate. Moreover, for a champion of the open economy, his position is incomprehensible. All the more so as, more clearly than elsewhere, he disarmingly writes that "[t]he 'natural' or equilibrium rate of interest which would exclude all demands for capital which exceed the real supply capital, is incapable of ascertainment, and, even if it were not, it would not be possible, in times of optimism, to prevent the growth of circulatory credit outside the banks" (Hayek, 1967, p. 125). The prudence of bankers in providing loans would be a solution because, right? not the one who asks is a madman, but the one who offers! Definitely, a difficult task! "[O]nly by a central monetary authority for the whole world" there may be some chances! (Hayek, 1967, p. 125).

Hayek's bank is very important in the management of forced saving. A bank that, however, does not control the situation and, unsettlingly, does not even want to! Faced with the inverted pyramid of the credit system, the central bank competes with commercial banks and non-banking institutions in multiplying lending. Nevertheless, they multiply what they receive from the source, from the Central Bank. There is the key. Hayek argues that the correct thing to do when the use of discounting becomes common practice and the production of derivatives reaches its peak is for the Central Bank to refrain from expanding loans and reduce their volume. Does this happen? No, he adds: "It is probably entirely utopian to expect anything of that kind from central banks so long as general opinion still believes that it is the duty of central banks to accommodate trade and to expand credit as the increasing demands of trade require" (Hayek, 1967, p. 117). Maybe! But what do we do? Do we notice the evil, declare the chance of its utopian cessation and surrender? And why does Hayek think that such an approach would be utopic? We believe that it is here, in his disarmament, the dead end of the refusal of the function of the standard of value of money. If he had not refused it, it would not have seemed utopian. On the contrary, it would not have seemed logical for the central bank to send money into circulation regardless of the size of the production. Hayek takes D.H. Robertson as his flag bearer. He quotes him in the motto of his fourth lecture: "The notion common (...) to 90 per cent, of the writings of monetary cranks is that every batch of goods is entitled to be born with a monetary label of equivalent value round its neck, and to carry it round its neck until it dies" (Hayek, 1967, p. 105). In his analysis, Robertson is not denied. Hayek is not one of the "cranks". He believes, just like Robertson, that the paths of goods and money can be parallel. And if this is the case, no matter the amount of goods on the market, we are allowed to send wagons of exchange means towards them. Which we keep, or not, until the very end! With such beliefs, you cannot claim that when the turning point occurs, when the money from an injection has done its "duty" and a reissue of the gesture is needed, the central bank will refrain. Beyond its profit interests, potentially growing to a large volume of loans, it is just doing its job, according to Hayek himself! If the growing needs of the economy require it, the Central Bank will deliver a larger volume of means of exchange! And so on! If you "scientifically" authorize it with such a mission, what else can you expect from it?! Can you write in its job description to take action in order to reach the ideal situation of a "constant rate of forced saving, or maintenance without the help of voluntary saving of capital accumulated by forced saving"? (Hayek, 1967, p. 151). Vain hopes with support in shaking sands on which he himself built the foundation of the theoretical construct. Theoretically, the urge not to find in "a forced credit expansion (...the) attempt to cure the evil by the very means which brought it about" (Hayek, 1933, p. 21) is justified. Practically, it remains empty and useless since you consider that it is the "duty of the central bank" to serve, as needed, the national economy with means of exchange! No matter how many!

Let us be clear here! Hayek, like all other Austrians, does not believe in the sincerity and goodwill of central banks. He did not produce any texts to encourage them to expand the economy with money wagons. He expects moderation and prudence from them. But he taught them a lesson that was too tempting: values are not important; the means of exchange do all the work! And the banks, in his hand, flood with means of exchange, which, it turns out, have nothing to do with values!

Trying, normatively, to suggest solutions, Hayek only complains. Credits to consumers in times of crisis equal failure; granted to producers, they could have a saving effect. They could, "if" and only if: ***if*** the volume of loans is adjusted to compensate for the initial and excessive increase in prices for consumers' goods; ***if*** the additional loans are withdrawn from circulation when the above-mentioned prices fall; ***if*** the ratio between the supply of consumer goods and intermediate goods is adapted to the demand for these goods. Moreover, Hayek concludes: "Frankly, I do not see how the banks can ever be in a position to keep credit within these limits" (Hayek, 1967, p. 98). Looking at the ratio between demand for consumers' goods and capital goods (Who should worry about this?) Hayek concludes, proverbially, returning to an "old truth" that "we may perhaps prevent a crisis by checking expansion in time, but that we can do nothing to get out of it before its natural end, once it has come" (Hayek, 1967, p. 99). To prevent this, we have seen it, we cannot count on the support of banks. Once the storm begins, we have to wait for it to be over. "so far our investigation has not produced a preventive for the recurrence of crises" (Hayek, 1967, p. 99). That is right! What do we have here? We have a colossal, yet purely Enlightenment intellectual effort, enlightening in the obscure and labyrinthic process of contradictory movements that define a cycle fuelled by forced savings. And we have another lesson about how we resiliently pay for the pleasure of living beyond possibilities. We understand, thanks to Hayek, what is going to happen if we dive in too much; if we live on dissaving, as Keynes put it. No one is guilty that "no measure which can be conceived in practice would be able entirely to suppress these fluctuations" (Hayek, 1933, p. 188). However, it does not deprive us of a ray of hope! Through public action, we could exert and avoid catastrophes. How do we avoid them? A suicidal thought is lurking around and, as if against his own intimate will, he writes: "...

certain kinds of State action, by causing a shift in demand from producers' goods to consumers' goods, may cause a continued shrinking of the capitalist structure of production, and therefore prolonged stagnation. This may be true of increased public expenditure in general or of particular forms of taxation or particular forms of public expenditure" (Hayek, 1967, p. 128).

We have here—we want it or not—the prescription for anti-crisis medication signed by Keynes. We tamper with demand by means of certain forms of taxation or certain types of government spending! In the works of Hayek we were looking for something else. And we came across a scholarly analysis, extremely labyrinthic, in a language that stands no chance of sending clear and compelling messages to the political class. Great promises with conclusions at the boundary between pessimism and uselessness, and with the urge that future analyses find a better place for clearing up "the relationship between interest rates, profits, and the liquidity of the banks" (Hayek, 1933, p. 238). With the suggestion that central banks be subjected to and exposed to a wider exercise of honesty and publicity, without this being an attack on liberalism, Hayek's lesson ends. A lesson in which we learn something about sustainability in one of the hottest areas of the economy, but we also find out that the chances of reaching it are modest. In brief, the recurrence of crises is related to the internal organic of the economic cycle. In 1987 Hayek was still alive. We do not know if he was aware of the Brundtland Report. Anyway, the Report's echo of the crisis did not give him a proper response.

1.2.2.3 Attempt of Improvement

It is a salutary and common fact that the same mind, in the dynamics of time, changes its thoughts. That the economy is a living construction site and those who serve it are aware of it is a well-known and accepted fact. Hayek is no exception to this rule.

In *Profits, Interest and Investment* he is tempted by such an ideational attempt; to return and revise some of the views expressed in *Prices and Production*. Through methodological corrections and changes of perspective, he honestly wants to tell us that he has reached other conclusions regarding the role of both the rate of interest and the rate of profit in the configuration and development of the cumulative process. If in *Prices and Production* the rate of interest enjoys an exclusivist privilege, this time the rate of profit acquires this status. In essence, Hayek is endeavoured by the attempt of a denomination of economic dynamics. In the obsessively burdensome relationship between the demand for consumers' goods and the demand for investment goods, the important role is allocated to the rate of profit, or the "profit schedule", as he names it. He seeks arguments to show that the trade cycle runs its course and that it ends up, inherently, in crisis even without the contribution of the rate of interest. In fact, he considers it inflexible, along with three other hypotheses, different from those in *Prices and Production* but just as artificial and unfriendly with an analysis demanding maximum clarity: the rigidity of money wages, limited mobility of labour and asset specificity.

Among all, the rigidity of the nominal rate of interest and that of money wages are the most "heroic" and act as the main arguments for initiating and supporting the analysis. The hypothesis of the rigidity of the rate of interest confuses him in clearly and correctly highlighting the role of the rate of profit, in a mechanism in which the gearing parts relate to each other, through inter-conditioning, in their movement and not their stiffness. The hypothesis of the rigidity of nominal wages makes the explanations of the Ricardo Effect—difficult to define anyway—nebulous.

The mechanism described by Hayek remains interesting and stimulating, a reason for reflection and intellectual pursuit; however, not so much in regard to the pluses that it claims to bring to *Prices and Production* or to what was already established as a gain in terms of analysis of the trade cycle, but, especially, through his well-known dilemmatic remarks and assumptions.

Correctly, the "scarcity of capital" nourishes a negative signal, and the recession is ascribed to declining investment and rising unemployment. Strangely, however, the decline in investment seems to be caused by the steadily rising rate of profit. A first critical point "does not depend on 'full employment' in general being reached, but on the capacity to increase the output of consumers' goods as fast as demand increases" (Hayek, 1975, p. 32). Another critical point of economic dynamics appears to be given by rising unemployment and declining incomes in industries whose activities take place in the earlier stages of production. Correct, but the hypothesis of the immobility of labour forbids this. Equally strange, Hayek operates with two equilibria. One, "a sort" of equilibrium, intended to be intelligible but remaining a mystery, is thought in terms of significant unemployment in earlier stages of production, high rates of profit for some and zero profit for others (or with profit below the interest rate!): "[t]he only thing which can bring this process to an end will be fall in employment in the second group of industries, preventing a further rise or causing an actual decline of incomes. And if labour is not mobile between the two groups of industries a sort equilibrium might ultimately be reached with a high rate of profit in the first group and no profits (or profits below the rate of interest) in the second group" (Hayek, 1975, pp. 28–29). Another one, a temporary quasi-equilibrium, is thought of in terms of high unemployment, equal demand and supply of consumers' goods and a stop to the shrinkage of production: "[i]n the end however, a new position of temporary quasi-equilibrium would be reached in which, with a very low general level of employment, the demand for consumers' good will once again have become equal to current output, and output and production will cease to shrink further" (Hayek, 1975, p. 36). Caught up in the mysteries of the "Ricardo Effect", Hayek is restrained by the paradox of saving. He rightly remarks, quoting Mill, that when income is saved, those who save "do not thereby annihilate their power of consumption; they do but transfer it from themselves to the labourers to whom they give employment" (Mill, 1885 apud Hayek, 1975, p. 44). Also relative to the destination of what is saved, the mention of Smith is commendable. J.B. Say who is not exploited here either would have been much more useful in providing and supporting the logic of transferring purchasing power from one group to another and in convincing that the savings belonging to a group can be used to increase the employment of another group of individuals in the capital goods industries. We believe that the assertion that the maximum level of employment is related to the difficulty of producing sufficient consumer goods is also limiting.

Emblematic for the nebulosity with which Hayek wants, but does not prove that he is able to bring more clarity to the ideas expressed in *Prices and Production*, is the emphasis on crisis as well as on the difficulty of finding a middle solution, between a too high profit level and a too low one. While, he believes, a too low rate of profit induces an increase in employment in both capital goods and consumers' goods industries in a proportion which corresponds to the natural distribution of labour—but which leads to a level of investment that can no longer be sustained by savings—a too high a rate of profit stops the investment process before the maximum and stable degree of employment is reached. A vexation, a dazzling dance between Scylla and Charybdis without a well-defined ending: "[a] policy designed to mitigate fluctuations will therefore have to watch the recovery to watch the recovery from its very beginning. The problem is to find a middle path between the Scylla of keeping the rate of profits too low and the Charybdis of keeping it too high" (Hayek, 1975, p. 61). The offered solution is just as difficult to digest.

As we have seen, the end of the road in *Prices and Production* leads to Keynes. Here too. Hayek does his own thing. The great universalist of liberalism remains impregnated by the neighbourhood of his former colleague at the London School of Economics. At the limit "[b]ut during the later half of the decline a policy of supplementing demand by public expenditure may well be justified" (Hayek, 1975, p. 63). Monetary expansion, to rapidly reduce unemployment, is not categorically excluded. Ultimately, as a solution driven by despair, it could be considered, but with caution because it perpetuates fluctuations.

Only by surmounting the hypothesis of complete immobility of the rate of interest, Hayek manages to outline the new architecture of the cumulative process according to the new premises and to tell us how to reach equilibrium and crisis. Allusively, it brings the interest rate mechanism closer to that described by Wicksell, considering that, in real life, no matter how tardily, the rate of interest follows the changes in the rate of profit and that the investment process is much more influenced by the rate of profit than by the rate of interest. At the end of the analysis, Hayek wonders whether the rate of profit must equal the rate of interest and thus reach an equilibrium. Here he assigns the natural label to the rate of interest, but not to highlight in a Wicksellian manner its virtues, but on the contrary, to undermine the illusory belief that something positive can be achieved for the trade cycle by manipulating the rate of interest as such. In his own terms, "manipulations of the rate of interest are of much more limited value than is often supposed because, if we try to fix it below or above its 'natural level', it soon ceases to be effective" (Hayek, 1975, p. 67). Altogether, Hayek is tormented by the fact that the empirical facts do not help him verify his thoughts, and thus, disarmed, he states that "[i]n particular I want to warn the reader that I do not mean to assert that the rate of profit actually does play quite the role which it is here assumed to play" (Hayek, 1975, p. 6). In fact, his sadness is rationally justifiable. The leading role of the rate of profit relative to the rate of interest cannot be emphasized in a mechanism in which the former was doomed to move around a fixed point—a frozen rate of interest as a courageous but disqualifying inoperable hypothesis in achieving the targeted objective—that of highlighting the role of prima donna of the rate of profit.

Overall, we believe Hayek wanted to achieve something else in *Profits, Interest and Investment* than in *Prices and Production*, but he did not fully accomplish his goal. His thought was daring and attractive. Aiming to cast the rate of profit and real wages as determinants of an economic dynamic shows realism and the intention to subscribe to sustainability. The rate of profit, as a guiding beacon, brings the economy closer to reality and subordinates the nominal to the real, according to a Ricardian and Wicksellian scheme. In Hayek's case, this idea is pallidly argued. As in *Prices and Production*, the hypotheses tangled him. Although he wanted to clarify and improve, initiating a new methodology, he remained nebulous, complicated, mysterious, and little convincing.

1.3 Concluding Remarks

If we do not pay proper attention, the world of money fills our science with mystery and weakens our judgements. This seems to be the conclusion of the above lines. And, we add, nothing seems more contradictory than to find that a track tailored on doctrinal reasons in the conception of money can be translated into a millstone tied to the neck of sustainability; that the lack of measure in manoeuvring it, bounderless "relaxation", induces the lack of aspiration in a dynamic with the virtues of resilience.

In more concrete terms, we want to say that Menger's doctrinal obstination defeated formal logic. His allergy to Marx meant not only the mass rejection of the objective theory of value, but also of the function of money as a standard of value. The history of the evolution of economic life and the analysis of this evolution tells us that it was not necessary to give up the standard of value. However, he did it, with no attempt to come back, and it was a source of visible and embarrassing misconceptions and breaks in analyses leading to misleading conclusions; attractive for the normative, for the political environment, but with an altering effect for the prestige of economic science.

It is suspicious to find that today there are very few those who notice how suspicious it is that all those who faithfully followed the command of the school leader now resume, from scratch, the thread of the demonstration so that through a tiring insistence they convince us that money is used to measure prices but they do not measure anything; that it is possible to establish value relations without having money as a standard of value. Neither Mises nor Hayek are reluctant to take over these Mengerian tautologies as such.

Cloistering in a system confines. Keynes did not get caught up in a system and, less liberal than the Austrians, was blunt in refusing the "barbaric relic". It hindered him in manipulating the rate of interest, which he linked, more than the Austrians, to the natural rate. He saw as available means the manoeuvrability of money supply, but not in any way and not however much; only to the point where Wicksell "said" Stop! What may seem contradictory, although it is not, is that his "green cheese" factory operates according to the logical scheme devised by Menger and perpetuated by Mises, Hayek and all contemporary Austrians. But who mentions Menger today when criticizing money loans out of nothing? Neither Mises nor Hayek are mentioned! The name we hear is Keynes. Total perversion! Anyone who criticizes the money loans out of nothing forgets that the theoretical "foundations" of this exercise are laid by the outstanding mentors of the Austrian school, mainly by Menger. They believe that money does not measure but only facilitates exchanges. Their theoretical anxiety about excess credit remains theoretical. Their marginalist dream is dictated by a commandment: prices and money have nothing to do with values!

On such grounds, we understand how Wicksell defeats the Keynes-Hayek dispute over the economic cycle. But he wins through Keynes, not Hayek. The first one leans on Wicksell, explains his work and relies on the natural rate in a broader context—that of development through saving, investment, consumption, but also of equilibrium through employment. As for the latter, he endeavoured to relativize or even to deny him; hindered to the point of annoyance by the general level of prices. The result is that—in times of crisis—Keynes is the claimed one. Even when the "liquidity trap" gives the measure of the enthusiasm of relaxation and says that a new injection is no longer necessary. Caught in the logic of the natural rate, Wicksell and Keynes put more value on interest rate manoeuvring than on monetary injection. Both the academic environment and especially the political one almost incorrectly attributes them the exclusivity of the chance in manoeuvring the money supply. This is not a victory for Keynes or Wicksell, but it is a defeat for *Menger and company*.

In the labyrinthic search for sustainability, Hayek proves to be a prisoner of his own contradictions. Weak premises and assumptions for the sake of the school "help" him build a methodology of unsustainability. Imaginary errors attributed to the standard theory, the chaotic movement of specific capitals between production stages, lectures for extra-terrestrial students who interpret graphs with four variables on two axes, etc., complete the Hayekian methodological arsenal.

Armed with such analytical tools, Hayek tackles the issue of the business cycle. The analyses of the relationship between the price and production structures, and the non-monetary causes of the business cycle—with many pluses and just as many minuses—are splendid lessons about something that does not happen. If somehow, however, it happens, it is not his recipe that inspires but a purely-Keynesian therapy. Because the end of the road, scholarly and dry to the point of headaches, ends just as dryly: "so far our investigation has not produced a preventive for the recurrence of crises". Our expectations are too high! If we are to find support in someone, he is not the right person. The pure ultraliberal sends us, subliminally yet clearly, to Keynes. The harshly criticized Keynes is recommended to be more useful through his concrete and undisguised exercise of honesty. Doctrinal equidistance puts him in a position to give a sign of good omen.

References

Bastiat, F. (2007). *The Bastiat collection*. Ludwig von Mises Institute.

Campagnolo, G. (2005). Carl menger's "money as measure of value." *History of Political Economy, 37*(2), 245–261.

Hayek, F. A. (1933). *Monetary theory and the trade cycle* (Vol. 1000). J. Cape.

Hayek, F. A. (1967). *Prices and production*. Augustus M. Kelley Publishers.

Hayek, F. A. (1975). *Profits, interest and investment and other essays on the theory of industrial fluctuations*. Augustus M. Kelley Publishers.

Keynes, J. M. (2013a). *A treatise on money. The pure theory of money*. Macmillan & Co.

Keynes, J. M. (2013b). *A treatise on money. The applied theory of money*. Macmillan & Co.

Keynes, J. M. (2018). *The general theory of employment, interest, and money*. Palgrave Macmillan.

Marshall, A. (2013). *Principles of economics*. Palgrave Macmillan.

Marx, K. (1990). *Capital* (Vol. 1). Penguin Classics.

Menger, C. (1892). La monnaie mesure de valeur. *Revue D'économie Politique, 6*(2), 159–175.

Menger, C. (2007). *Principles of economics*. Ludwig von Mises Institute.

Mill, J. S. (1885). *Principles of political economy*. D. Appleton And Company.

Mises, L. (1998). *Human action*. The Ludwig von Mises Institute.

Mises, L. (2013). *The theory of money and credit*. Skyhorse Publishing Inc.

Ricardo, D. (2001). *On the principles of political economy and taxation*. Batoche Books.
Salin, P. (2020). *Tax Tyranny*. Edward Elgar.
Say, J. B. (1971). *A treatise on political economy*. Augustus M. Kelley Publishers.
Smith, A. (1977). *An inquiry into the nature and causes of the wealth of nations*. University of Chicago Press.
Wicksell, K. (1962). *Interest and prices. A study of the causes regulating the value of money*. Sentry Press.

CHAPTER 2

In Search of a Lost Lesson

We are thinking about a lesson in the registers of which policy can find the resources it needs to avoid major crises. Like any science worth its salt, economics should also own it; in a textbook signed by outstanding authors. Well, it does not. The great ideas on the subject are centrifugal and their authors are in constant and "innovative" opposition. Overall, while the nominal economy exploits the real economy, there is chaos in the theoretical registers. We do not have enough years to figure it out—Mason and Butos (1996) are convinced. Yet, sustainability needs solid theoretical support. We looked for this support in the works of scholars like Menger, Mises, Hayek or Keynes. We believed they, more than others, were following in the footsteps of the classicists. It turned out their offer did not fully live up to expectations. That is why we are looking elsewhere. Therefore, we are looking for strong ideas that are likely to inspire; and we are wearing out our boldness on the following landings.

2.1 Exploring the Outskirts of the Business Cycle Sustainability

Here, on the issue of the business cycle, economic science should be very close to the positive; it should be technical, rigorous and uninterpretable. However, it is not hard to notice that this is not the case. If we are willing

I. Pohoaţă et al., *The Sustainable Development Theory: A Critical Approach, Volume 2*, Palgrave Studies in Sustainability, Environment and Macroeconomics, https://doi.org/10.1007/978-3-030-61322-8_2

to see that we have different—even opposing—views on the triggering factors, the role of money, the current picture, the specific weight and importance of some parts of the cyclical mechanism (such as employment, the role of the bank or government, of the free market, the primacy of some moments to the detriment of others, conclusive solutions, etc.), we realize that we do not speak the same language and, as such, we cannot provide solid and inspiring theoretical registers for a reassuring mechanism of the healthy and resilient functioning of economy.

We find the placement of ideas and their authors on theoretical-doctrinal orientations to be productive insofar as the competition comes up with indisputable essences instead of conclusions. The gain would be considerable if here, in the area of large syntheses, there were points of interference and contact areas produced under a great name, closer to the logic of things than to the interested philosophy of a school or system. Although we are aware of the fact that the niche regarding the points of contact is not so generous, we guide our searches in the same direction. We take the path of the outstanding names. It is natural to look for possible analysis affinities in those involved in the contemporary ideational competition on the subject. Friedman, Fisher, Minsky, Rothbard, Huerta de Soto, Krugman, Stiglitz, Roubini, Greenspan and many others allow us to exploit their works. What do we learn from them in support of sustainability? Broadly speaking, we discover diversity in places that demand unity and principles; puzzling places instead of clear sources of inspiration for policy; few areas of interference allowing chances of outlining a resiliently valid paradigm.

2.1.1 "Heresies" of the New Austrians on the Topic of Business Cycle Sustainability

We would have expected that the theoretical-doctrinal scope of Mises or Hayek—with all its logic contortions and breaks—to find completion attempts, more doctrinally relaxed and useful for both politics and science. But Rothbard, de Soto, Hülsmann, Hoppe, Block and others "are specializing". They provide an enticing analysis, it is true, in the area of the business cycle, trained and supported by and with money. However, we barely hear about the natural interest rate, scarcely represented by their spiritual parents. Hülsmann (2008) mentions it briefly. In return, we have got from them refined and profound arguments about the inherently unsustainable mechanism of money out of nothing, about the uncontrollable lending mechanism of the double or triple inverted pyramid and, why not, about the stock market alike game subscribed to countless risks or fraudulent insurance.

In the heading *starting point*—we find out about a proven truth, as a removable yet unremoved evil: the unsupported warehouse receipts issued by banks and their nebulous movement are underlying causes of the boom and recession recurrence through the "expansions and contractions of the counterfeit bank credit" (Rothbard, 1994, p. 40). Freshly printed money can be fake warehouse receipts, with no coverage but formally identical to the genuine ones, unlawfully borrowed. The "Austrian" bank has the initiative on its side; it becomes a fraudulent entrepreneur who sets in motion a parasitic mechanism doomed to failure. As they are counterfeit or less counterfeit, more receipts equals more money. Rising prices, reducing the purchasing power and redistributing income make up for one effect. The loss of reputation—the compulsory prefacing of the qualified thief status—constitutes another effect. But that's not all. The allowed compromising system of fractional reserve allows the production of credit money without problems. If this "new money" reaches companies first it produces the terrible and undetectable phenomenon of the economic cycle. Counterfeit additional money, unrelated to voluntary savings, induces a sense of a propensity to invest. Higher prices on newly attracted factors or higher salaries will be paid. As new money is metabolized, the consumption—savings natural ratio is restored, and the error of unprofitable investments is revealed. Their liquidation triggers the depression phase of the economic cycle (Rothbard, 1994). Who is to blame for uninspired investments? The bank that generates an impulse based on counterfeit, or the entrepreneur for letting himself be misled?! Neither, Hayek would say. The system is to blame because it is legally open to unlawful actions. And the system means rules; more or less unstable but made by "masters" for "masters". This is why Huerta de Soto spends an impressive number of pages in his book *Money, Bank Credit, and Economic Cycles* (de Soto, 2006) on the subject of legislation in economics and its outstanding role and importance.

Through Rothbard, the Austrians also imagine an idyllic course of the cycle. The premise is a stable currency, preferably gold or silver. Under such conditions, "in a progressing economy, the increased annual production of goods will more than offset the gradual increase in the money stock. The result will be a gradual fall in the price level, an increase in the purchasing power of the currency unit or gold ounce, year after year. The gently falling price level will mean a steady annual rise in the purchasing power (…), encouraging the saving of money and investment in future production. A rising output and falling price level signifies a steady increase in the standard of living for each person in society" (Rothbard, 1994, pp. 20–21). Said, not done. The picture is idyllic. That is what is desirable! Until we say why it is not real, we need to acknowledge the "dissident" terms and the philosophy compared to that of the great Austrian masters. We remember what they thought about the purchasing power, about the calculations in terms of labour costs and productivity, about maintaining the level of nominal wages and increasing real wages. What Rothbard says above is similar to what Hayek said on page 105 of *Prices and Production*. Both infer chances for economic welfare from lagging behind the amount of money in relation to the increase in production—disproportion resulting in a price cut. From here, Hayek's economic welfare means a gain in productivity; that of Rothbard equals an increase in the purchasing power of money, with open opportunities for future savings and investment. Within this context, Hayek at least entertained Marshall's idea of the price-productivity relationship. Rothbard no longer needs Marshall, nor does he mention him. The judgements of the former resonate with those of the classicists. From there we learn, in a healthy way, that low prices are a result of reduced costs, which in turn is caused by increasing productivity. This is also the classic way of making profit. Had he remained true to the Austrian parents' belief that it was not the costs that determined the prices, but vice versa, Rothbard would have stumbled and failed to reach this judgement. "Dissidence" helped him, for a moment, to think like a true classic.

Hülsmann, a representative of contemporary Austrian doctrine, does not forgive him for the statements above. In *The Ethics of Money Production*, fighting like a firedrake with the "seven most widespread errors" of the standard monetary theory, he declares that he is angry at the sophism and the "the naïve belief that economic growth is possible only to the extent that it is accompanied by a corresponding growth of the money supply" (Hülsmann, 2008, p. 60). The young Austrian takes Say and Mises by his side and puts John Law, Wieser and Schumpeter at gunpoint. It is true that Rothbard did not mention, in the lines above, a *directly proportional* relationship between production and money. But he "gave himself away" and looked at them hand in hand. Or, Hülsmann is bothered by any connection. For the simple reason that "any quantity of goods and services can be exchanged with virtually any money supply" (Hülsmann, 2008, p. 61). If, somehow, this latest version of the truth would lead entrepreneurs to bankruptcy, forcing them to sell their products at prices below cost, Hülsmann comes to the rescue. He equips them with rational "expectations"—as we deduce from the context—based on which they can reduce their costs by being able to make a profit even when they sell at a falling price. This, he believes, would be a state of "normality", not a phantasmagorical hypothesis. And to convince us, Hülsmann refers—to whom do you believe?—to Milton Friedman and Ana Schwartz with their *A Monetary History of the United States*. Therefore, criticizing an "error" in which, unwittingly, you included an ideational brother, you end up leaning on a titan whose shadow you run away from but, at the same time, you are forced to find that you cannot be more than he wants you to be!

Perplexity is all the more significant as Rothbard, confrère of ideas with Hülsmann, declares his categorical disagreement with Friedman. The occasion is related to the cyclical route model dreamed of by the economist of the Chicago School. Rothbard believes that the "temptation to counterfeit" money and filling in circulation channels with this money does not occur according to Friedman's "helicopter effect" version but is based on the "ripple effect". The monetary surplus, Rothbard argues, is not distributed to every economic player proportionally to what they already have, as Friedman's monetarists believe, thus leading to a simultaneous increase in all prices. No, instead of a "magical and equiproportionate expansion", money follows at least a two-stage process. First, those who multiplied the money benefit from it. With the money surplus, they purchase goods. Their growing demand leads to an increase in prices. Those who have not yet received the money realize that they are at loss; they "find the prices of the goods they buy have gone up, while their own selling prices or incomes have not risen" (Rothbard, 1994, p. 24). While the first to take possession of the money win, those still on the road or at the end of this chain—causally imposed—lose. The conclusion—strongly Wicksellian—is very important: if there is a time lag between the infusion of money and rising prices (relative and by no means general) the reduction in money supply does not lead to an immediate cessation of the effects of monetary "injection" in prices, in the distribution of income, expenditure or production—as some supporters of Friedman' quantitativism might believe. A "big error"—Rothbard rightly believes—for which all the supporters of the Chicago School have fallen.

Overall, while Rothbard rightly criticizes Friedman, Hülsmann, who belongs to the same ideational space, relies on his arguments to impose his idea. What a lesson on sustainability; ready to be submitted for reading to any government interested in coherent analyses and conclusions; prefaced, if necessary, by the "most widespread errors" bearing Hülsmann's signature: (1) "the absolute money supply of an economy is virtually irrelevant" (Hülsmann, 2008, p. 63); (2) "hoarding (…) is never a monetary problem" (Hülsmann, 2008, p. 64); (3) "from the aggregate (social) point of view, it does not matter who controls the existing resources" (Hülsmann, 2008, p. 67); (4) "it is by no means sure that politically induced increases of the money supply will lead to a decrease of the interest rate below the level it would have reached in a free economy" (Hülsmann, 2008, p. 72); (5) "[s]tability of the purchasing power of money does not at all come into play" (Hülsmann, 2008, p. 79); (6) "commodity monies such as gold and silver feature a *built-in natural insurance* against an excessively depreciating purchasing power of money" (Hülsmann, 2008, p. 79). We could be accused of taking ideas out of the context that generated them. We are willing to take this chance. Each of the six sentences mentioned deserves a critical essay. We will stop here only on the last two: belonging to two successive pages, they contradict each other, obviously and disarmingly. In the absence of a well-deserved comment on each of the six sentences, we allow ourselves only one question: can a theoretical construction formed by the six "bricks" stir the admiration of some economic and social constructors looking for inspiring scientific treasures? We doubt it!

2.1.2 On the Sustainability of the "Monetarist" Cycle

Who can fit into the Austrian range of ideas—appreciative or critical—on the subject of the economic cycle? A first answer, but not the only one, refers to Milton Friedman and his monetary quantitativism. Just like the Austrians, Friedman sees in the business cycle a mirror of the money cycle (Friedman, 1993). Like them, whom he mentions very rarely—only when he imputes something to them—Friedman considers that the monetary impulse is everything in the economic dynamics. But unlike them, Friedman believes that in the long run there are chances for economic stability; that is, the economic mechanism is not endemically destabilizing. Just like the Austrians, Friedman believes that money created out of nothing and allocated for government spending fuels inflation. Unlike them (especially unlike Mises), Friedman considers "to be no systematic connection between the size of an expansion and of the succeeding contraction (...) a large contraction in output tends to be followed on the average by a large business expansion; a mild contraction, by a mild expansion" (Friedman, 1993, pp. 171–172). Friedman's conclusion is inferred from a framework of analysis that the Austrians reject; it is the frame of Robertson's "fanatics". Working precisely in this perimeter, working as a "fanatic", Friedman restores a relationship—fundamental both in his work and in that of Ricardo—between the production volume and the total amount of money. He believes and convincingly argues that an upward or downward movement of economic activity is followed by a movement of money in the same direction, although not strictly proportional. There might be another area with possible comparisons in which some Austrians have unfairly criticized him. It aims at the systematic process of the economic mechanism triggered by a new infusion of money. According to Friedman (1970) a monetary injection takes six to nine months to find a rise in nominal income and a similar interval until it finds a rise in prices. That is, on average, the extra money can lead to inflation in about 12–18 months. This allows us to infer that you cannot stop, on a short leash, an inflation that has already started. So where did Rothbard get the "helicopter effect" that he attributed to Friedman? We see that the propagation is also done on a ripple-like basis, gradually! On top of that, in the "Austrian spirit", Friedman is convinced that the above-mentioned relationship needs time in order to be perceived; "is not obvious to the naked eye" (Friedman, 1970, p. 10).

As far as the intimate processuality of this mechanism is concerned, however, there are more interference points between Friedman and Keynes. Similarly to the author of the *General Theory* (Keynes, 2018), he finds that a monetary impulse has the "liquidity effect", an effect reflected in balance sheets and not in income (Friedman, 1970), with real chances to reduce the interest rate. An interrelationship between economic players explains why, in a next sequence, what for some constitutes an expense, for others means income. There is a dispersal of money, and thus the effect on the balance sheets turns into an effect on income and expenditure. "But this is only the beginning of the process not the end [he warns]. The more rapid rate of monetary growth will stimulate spending, both through the impact on investment of lower market interest rates and through the impact on other spending and thereby relative prices of higher cash balances than are desired. But one man's spending is another man's income. Rising income will raise the liquidity preference schedule and the demand for loans; it may also raise prices, which would reduce the real quantity of money. These three effects will reverse the initial downward pressure on interest rates fairly promptly, say, in something less than a year. Together they will tend, after a somewhat longer interval, say, a year or two, to return interest rates to the level they would otherwise have had. Indeed, given the tendency for the economy to overreact, they are highly likely to raise interest rates temporarily beyond that level, setting in motion a cyclical adjustment process" (Friedman, 1968, p. 6).

What do we see here? A lot of Wicksell! We also see that the shadow of money as the equivalent of value and price as an expression of value preoccupies Friedman. Otherwise, the wording "raise prices, which would reduce the real quantity of money" would not be supported. We also notice something important in the long quote, reproduced as a whole because it expresses a summary of the vision of the cycle from Friedman's perspective: we notice that within the cycle, the interest rate changes its meaning. First it decreases, then, after the extra money becomes income—as Wicksell thought, it begins to grow. Therefore, Friedman believes, based on his own observations, the interest rate is a "misleading indicator" for the decision-maker. If we can focus on something, that something is the money supply, not the interest rate. At the same time, Friedman does not miss the opportunity to tell us that large variations in the money supply are not good either; on the contrary, they are destabilizing and should be avoided. Opposing a discretionary monetary policy, he recommends—deeming it sustainable—a constant increase in the money supply, by 4–5% annually (Friedman, 1970). And, as a paradox, which he also allows himself based on his own analyses and observations, Friedman does not rule out a monetary authority to ensure low nominal interest rates, engaging in an anti-deflationary policy and, conversely, to ensure high interest rates through an anti-inflationary policy. This is in close connection to changing the direction of interest rate within the cycle and starting from the premise that this is a result (not an a priori usable tool) of money management. Which option leads more to sustainability?

On the issue of the anatomy and dynamics of the economic cycle, Friedman is interesting from other points of view as well. His methodological inventions "permanent income" and vital cycle of economy place his conclusions on a different paradigm from that of Keynes. And this, considering the fact that the leader of the Chicago School literally borrows and exploits, much more than the neo-Keynesians, the Keynesian idea of anticipations. Relegated to the consumer, this premise induces the idea of "permanent income". Friedman's consumer is able to anticipate and calculate. He builds a life plan based on what is constant (permanent) in his income. If we start from these individual life plans, the evolution of the economy is not only predictable but, he believes, is less subject to cyclical fluctuations than we think. A widespread indexation of prices, wages and interest rates to shield the money from the turmoil of the economy would further strengthen the belief in the long-term stability of the economy.

If we can speak of a life plan induced by an autonomous "permanent income", conceived in a currency that, in the long run, is assimilated by Friedman to a "superior asset", then is the Keynesian function of the preference for liquidity still valid? The study of the US monetary history for nearly a hundred years leads him and Ana Schwartz (1963) to notice a growing trend in the individuals' and the community's preference to hold liquidity as real income increases. Here, money turns into a "superior asset". In other words, the money supplement does not satisfy the "hunger for liquidity" so as to determine, beyond this point, a decrease in the rate of interest; on the contrary. There is one more reason for Friedman to believe that we cannot rely on the rate of interest as a strong instrument of economic policy. The liquidity trap is proof enough for him.

The same idea of Keynesian anticipations is used by Friedman to explain the inflation-unemployment relationship. Here, he also wants to be a Wicksellian. He borrows from Wicksell the scheme of the relationship between the natural rate and the market rate but transposes it into the field of the relationship between inflation and employment. He undertakes this operation to argue that we can only fight inflation in the short term; that there is a "natural rate" of unemployment compatible with any level of inflation. Moreover, the inflationary boom may lead to an increase in employment. However, the phenomenon of anticipation results in a return of unemployment to the previous level but with higher inflation. Here is what Friedman textually says: "Employees will start to reckon on rising prices of the things they buy and to demand higher nominal wages for the future. 'Market' unemployment is below the 'natural' level. Even though the higher rate of monetary growth continues, the rise in real wages will reverse the decline in unemployment, and then lead to a rise, which will tend to return unemployment to its former level. (...) there is always a *temporary* trade-off between inflation and unemployment; there is no *permanent* trade-off" (Friedman, 1968, p. 10–11, our emphasis).

So what's to be done? If we return to the natural unemployment rate at the cost of higher inflation, is this worth the try? Yes, Friedman will say, along with his colleague Edmund Phelps! On one condition: generalized indexation of revenues in relation to prices! And thus, does the economy become more resilient? Hard to say!

The above ideas bear Friedman's signature, but also share neighbouring sources. He made a coagulation and draining effort by recalling the names of many economists from whom he got inspired, to wit, K. Bruner, K. Wicksell, *A Treatise on Money* of young Keynes, F. Knight, T. C. Koopmans, M. W. Holtrop, J. Rueff, A. Meltzer, J. F. Muth, R. E. Lucas, R. A. Mundell, A. C. Pigou and many others. However, he followed Fisher the closest.

How does Friedman get our attention, what emphasizes the essence of his thinking? The long term vs. the short term sends him to the necessary scientific compromise. In this pursuit, he is unwilling to sacrifice the economic real. We modify income, indexation and prices. Every effort is worth it in order to avoid unemployment. This is the worst and it "hurts" directly. Unemployed, we are not entitled to talk about sustainability.

2.1.3 Irving Fisher: Over-Indebtedness—Deflation—Monetary Injection. And if You Are Poor, Chances Are You Remain So!

It is no coincidence that Fisher's work came to Friedman's attention. The author of *The Theory of Interest* (1930) and *The Debt-Deflation Theory of Great Depressions* (1933) innovated in the field of quantitative money theory. He was also concerned, just like Friedman, with the relationship between the quantity of money and the level of prices. Its famous equation $MV = PT$ made history. Treated and analysed from all doctrinal positions and directions, the alleged causal link between the variables of the equation with the rank of synthesis of quantitativism proves, at the same time, that it expresses a great truth but also a splendid tautology. Fisher himself returned to his own equation but the conclusion did not change it. Not even Friedman—in a new formulation, $M = K \times P \times J$—changes the basic idea: the only variable parameters remain the money supply and the prices and the movement direction of the causal relation is from M to P. The other parameters are frozen. Moreover, not even the Cambridge version, through Marshall and Pigou, confined to the micro-level, is more than a mere complement to the "transactional" one. However, introducing Keynes in the dialogue, Friedman borrows the observation regarding the "V" element in the formula, the velocity of money circulation. This cannot be constant. On the contrary, it increases or decreases as M decreases or increases. The high-flow river flows slowly, Cantillon said. And if that is the case, reducing the velocity of circulation could nullify the effects of the quantitative increase on the price level. A "trap" could turn a new injection into water off a duck's back. In addition, velocity can change depending on income. Thus, M, the amount of money does not seem so important to Keynes. Truly important are investments and government spending (Friedman, 1968).

Fisher is worth mentioning, in context, for *two* other important ideas. The first is a credit-induced division of the world. The second refers to indebtedness and unindebtedness.

1. The author of *The Theory of Interest* does not seem to be seduced by the idea of a "permanent income" under any circumstances. Interest on loans will sanction the future purchasing power. It is possible for income to increase but the purchasing power to decrease because of debt. The rich can overcome this discomfort, as they can save and accumulate. Things are different for the poor. They will mostly be concerned with the present; a present in which they need to acquire goods, with little or no chance of saving and escaping poverty in the future. Fisher's conclusion is definitive: "the great masses, once they get near the bottom, are likely to remain there" (Fisher, 1930, pp. 339–340). Mobility on the scale of the social hierarchy assumed by A. Smith, depending on the way, more or less efficiently, in which everyone spends their energy, remains a beautiful dream in Fisher's eyes.
2. Indebtedness and unindebtedness shape the broad outlines of Fisher's economic cycle. It is absurd to believe, he convinces us, that after long searches, the economy finds equilibrium and remains this way! (Fisher, 1933). No, the economy is in a constant quest towards an upward trend. Fisher rejects the traditional approach in detecting the causes of a crisis with an emphasis on: overproduction, under-consumption, savings-investment gap, etc. For him, the ultimate cause is over-indebtedness followed by deflation. Following the sequential succession of the causal chain ending in crisis, it should be noted that: (a) Relaxed lending equals investment opportunities; (b) All goes well until the revenue from payment collection becomes insufficient for the payment of loans (principal + interest)

and you have to contract a new loan to pay your outstanding debts; (c) You do not resort to a new and burdensome loan; you give up a new loan and start selling assets to settle the accumulated debts; the consequences of such unindebtedness are: reducing deposits, prices, income, the market value of companies and employment. Overall, contraction and depression! (Fisher, 1933). To get rid of the whole procession of fatal events caused by deflation, Fisher comes up with a solution—the same that Friedman will suggest—that is, to stimulate prices by manipulating the money supply. In his terms, deflation implies resuming lending, in order to allow income consolidation and, thus, the payment of outstanding debts (Fisher, 1933).

What do we have here? The idea of reinjection with additional credit comes up in Fisher's writings not as a solution that leads to fatality; on the contrary, as a chance of revitalization strictly necessary to get out of the debt crisis. To him, the relationship between receipts and debts is very important. Moments of imbalance between the two terms of the mentioned relationship appear as long as it is not possible to do without credit. And, basically, what does Fisher want to tell us? He is telling us that supply is not always able to properly produce demand to the extent necessary to ensure an upward trend in economic dynamics. Not only do you have to borrow in order to grow, in economic terms, but you also end up borrowing along the way, for possible but likely temporal and amplitude discrepancies between the time and size of the receipts and debts. Say is hinted at here. Fisher does not accuse him by mentioning him. Keynes will surely do this. For him, the main cause of the evil called crisis resides in the danger of having the actual demand falling behind!

2.2 Renowned Interpreters in Search of a Sustainability Deprived of Real Sources

The belief that by acknowledging or criticizing a school or an outstanding scholar, you borrow something from their aura is well known. Economy is no exception. Reputable interpreters quarrel (!) with Hayek, Keynes, Friedman or Marx, with the classical or neoclassical school either to increase the share of their own ideas or to bring an addition to the history of economic analysis. In our field, many are those who look for sustainable ideas in the yard of established schools or scholars. We will stop at a few (it is difficult to define representativeness here) to prove how unfriendly the theoretical offer is, even at prominent scholars, and how necessary a clear lesson on sustainability is.

2.2.1 *Mason and Butos: Keynesianism Is Sustainable, Monetarism Not at All!*

An attempt to prove that we do not always find sustainability where we are looking for it and that it is not easy to produce syntheses between schools is provided by Will E. Mason and William N. Butos, the first author, the second editor of the work *Classical Versus Neoclassical Monetary Theories: The Roots, Ruts, and Resilience of Monetarism—And Keynesianism*. We stopped on this work due to the large number of references in the area we are interested in. Through the title, the authors invite us to a journey among the ideas of the classicists and neoclassicists. This is just an opportunity. Against this background, two great contemporary schools are compared, revealing—as the authors argue—serious issues in the field of monetary theory. These are Keynesianism, with its newer versions, and monetarism, led by Friedman and the Chicago School.

The text of the mentioned work is an attack at Friedman's monetarism with all its "sons" and sympathizers, seen and unseen, accusing it of falsity and absurdities. As interpreters of the classical doctrine, the monetarists, along with Fisher, Kemmerer, Hicks, Patinkin and Schumpeter, have lost sight—the authors tell us—of the transitory nature of the quantitative theory of money, judging it as a long-term doctrine, considering that the classicists abstracted the "clock-calendar time" (Mason & Butos, 1996, p. 34). The work of Pigou reveals that the latter "failed to recognize that bank deposits had, in fact, become money" (Mason & Butos, 1996, p. 43). The case of Fisher is just as bad! "[T]he equation of exchange is not an actual equation, but only a shorthand method of arranging the variables in a manner convenient for analysis" (Mason & Butos, 1996, p. 49). Moreover, "Pigou's 'cash-balance' turned out, in effect, to be, simply, the reciprocal of Fisher's 'velocity'" (Mason & Butos, 1996, p. 43). Still in the lens, Milton Friedman, a sort of Hamlet, is believed to be "charged with the responsibility for coming up with the appropriate supply of money on cue (...) rendering the post-World War II inflation as apparently incurable as it is incomprehensible" (Mason & Butos, 1996, p. 50). And, after all this, here is a load of sentences through which the authors, in search of sustainability, send us to chaos:

"the notion of money as a veil that does not affect the real economy (…) originated as a pseudo-classical distortion of classical doctrine" (Mason & Butos, 1996, p. 78); "Unlike the classical quantity theory, the neoclassical 'quantity theory' could not account for the 'quantity' of money" (Mason & Butos, 1996, p. 80); "Patinkin based his critique of neoclassical monetary theory on Oscar Lange's misconstruction of classical theory" (Mason & Butos, 1996, p. 81); "Lange's crucial statement failed to make a distinction vital to classical and neoclassical monetary analysis namely, the differentiation of 'commodity money' (specie) from the 'money commodity' (gold)" (Mason & Butos, 1996, p. 84); "Lange's analysis appears to have been merely nominal application of 'Say's Identity' to a monetary economy" (Mason & Butos, 1996, p. 85)!

Who can handle this "sustainable warp"? It seems that neither those who photograph it. A serious accusation draws serious attention: "monetarists appear to have consciously removed the quantity and value of money from the jurisdiction of monetary theory" (Mason & Butos, 1996, p. 88) and "Central banks, which have generally adopted the monetarist research program (…) to hide their own inability or unwillingness to control the money supply" (Mason & Butos, 1996, p. 89). Is there anything else necessary to complete the feeling of chaos? Here: Paul Volcker admits and supports a False when he believes that "targeting monetary policy on a fixed growth rate for some 'monetary aggregate' tends to be stabilizing for the economy" (Mason & Butos, 1996, p. 97). The cyclical mechanism derived from monetarism based on the risky game of the free market appears to the authors as degenerate, speculative, "money-making" rather than producer of goods. The relatively stable long-term economy is an "irony" and Friedman's rational expectations are "absurd". What else are the monetarists guilty of? They are guilty of not having realized that two phenomena occurring simultaneously (considering the incremental amount of money and the investment) decline the causal approach! Incidentally, we know that Friedman admits the lag in question, just as the "absurd" rational expectations can equally be found in Keynes's welcoming yard.

Overall, there is no good word for the Chicago monetarists! If they and the central banks that find inspiration in their works have tried the inability "to decide whether the interest rate is an instrument (means) or objective (end), i.e., an 'indicator' or 'target' (in monetarist terms), of monetary policy" (Mason & Butos, 1996, p. 107), what else can we claim? Friedman's successors update and build a deformed, disarticulated and, above all, dangerous product. Or they lean on both the Chicago school and on rational expectations, indulgent to the FED relaxation policy, just as R. Lucas does!

Instead, for Keynes and neo-Keynesians, Mason and Butos have nothing but words of praise. Their expectations do not smell bad! Keynes opened roads and theoretical and practical perspectives. He "described 'the way in which changes in the quantity of money affect prices in the short period'" (Mason & Butos, 1996, p. 151). Then, Keynes is more classical than the so-called neoclassical theorists because his "theory of the value of money, by contrast, together with the medium of exchange conception of money, restores a unity of value theory analogous to that of the classicists, allowing a return of monetary theory to classical methodology" (Mason & Butos, 1996, p. 156). A strong point comes from the possible Keynes-Hayek association on the line of methodological subjectivism and, on a very important place, that of the optimal amount of money! [Idem] which only the market can discover and where Hayek and Keynes think alike, as Mason and Butos argue.

We could continue with the obvious or "discovered" strengths of Keynesianism. On the whole, neoclassicism and its contemporary extension in the guise of monetarism, plus the central banks it feeds on ideas and which have illegally expanded their "independence", refuse or do not know how to define money but aim to manage it. Instead, the neo-keynesians know them all and do them all well. They do not forget that "fiscal policy is a *species of* monetary policy" (Mason & Butos, 1996, p. 172); they note that the interest rate is "a lagging stabilization instrument" (Mason & Butos, 1996, p. 182); and, very importantly, the Keynesians realize that Laffer's doctrine of supply "was the legitimate reaction against the quasi-monetarist Fed's willingness to sacrifice production and employment to the Procrustean monetarist remedy for inflation" (Mason & Butos, 1996, p. 180).

We learn, *conclusively*, thanks to the two authors mentioned, that the monetary theory in the two versions can be just as solid and just as groundless. Who should we rely on? On the two authors, who, disarmingly, calm us down by announcing that "[i]f it takes as long to sort out the pseudo-, neo-, and new- branches of Keynesianism as it has taken to make the equivalent differentiation of classicism, none of us will live to see these issues resolved" (Mason & Butos, 1996, p. 77)? It is in the spirit of the philosophy of sustainable development to pass on to the next generation a dowry, clearly outlined. What a chaos do we pass them on!

2.2.2 *The "Minsky Moment": An Erudite Analysis Lacking Sustainable Message*

But what do we learn from the "Minsky moment"? His much-cited work *Stabilizing an Unstable Economy* promises, right from the title, a magic recipe. There is cure for a balky economy—and, on top of it, capitalist—doomed endemically to instability; there is a recipe which, if respected, leads to stabilization and, we hope, deceived by the author, to sustainability and resilience.

We say from the beginning that, although the title is very seductive, the book is disarmingly unfriendly with real and credible solutions. It is an interesting analysis, covering hundreds of pages and aimed at promoting lesser known places, "allergic" to the classicists and the Austrians, which does not coagulate in strengths in order to come up with solutions worthy of an imitable therapy. The picture—to a lesser extent the film—of the targeted places is excellent. The steps to be followed are either faintly marked or trapped in a paradigm with little support in the intrinsic rationality of the economic mechanism.

We suspect that the alleged recommendation of the existence of the "Minsky moment" is related to the accuracy with which he photographed the phenomenon of speculative financing in recent decades; a definite cause of economic instability that is difficult to manage. This instability is not only inherent and fundamental, but, as he correctly warns, the world "subscribes" to it, getting used to the evil. The final acceptance of risky practices, unsanctioned but validated by the authorities, turns a small evil into an asset for accepting a greater evil in the future.

The main cause of economic instability, starting with the 1960s, is the result of "[t]he weakness of the banking and financial system" (Minsky, 2008, p. 57). What does this process mean? Nothing more, nothing less than the expansion, outside the central bank's control, of derivative products resulting from the relations of commercial banks. Following the example of the FED, the mechanism is clearly described by Minsky. The lending rate of the central bank influences the behaviour of private banks as far as their financing operations are concerned. This until "bank reserves are mainly the result of open-market purchases of government securities" (Minsky, 2008, p. 282). From here on, the central bank loses its influence, the "daughter" banks go crazy, borrow from each other and finance themselves speculatively, ignoring the "canons of good business practice". The distortion of these good practices is accomplished by decentralizing the main function of the central bank: by delegating authority, giant commercial banks become, in turn, lenders of last resort, ready to break any kind of canons. And they break them by: (a) producing "financial innovations" at a rate that exceeds the restrictive intentions and practices to which they are subject to; (b) borrowing from one another in order to pay their debts and inventing "non-banking" funding sources; (c) mixing their portfolio to such an extent that they no longer know if they are commercial banks or investment banks.

Minsky's contribution to capturing the image of the cancer-causing mechanism fostering the proliferation of banking products that we call toxic is notable. However, we also believe that other names deserve to be added to his "moment". For instance, although not mentioned here, Hayek explained the phenomenon clearly and comprehensively in 1931, in *Prices and Production*. The inverted pyramid said that on the wider side of the pyramid, private banks escape the control of the central bank. A disarming fact both for him and for Minsky. Rothbard is equally entitled to be associated with the "moment". Interested in the pyramid pattern of credit expansion, the dubious honesty of commercial banks, the fraudulent game of counterfeit receipts and the inherent inflationary nature of fractional reserve credit systems, he put forward a splendidly clear analysis in 1963. See again *What Has Government Done to Our Money?*

There are other sources of the much-cited "moment" in relation to which we cannot attribute to Minsky a whole scientific adventure novel. He himself declaratively states that "[t]he fundamentals of a theory of financial instability can be derived from Keynes's General Theory, Irving Fisher's description of a debt deflation, and the writings of Henry Simons" (Minsky, 2008, p. 192). Don Patinkin's name is not forgotten either. He is caught up in Patinkin's solution which, by the way, is no novelty. It is an explanation in other words and deliberately twisted of the mystery of demand: extra demand is needed for extra money supply to find reason! How clearly others had said it before Patinkin and also Minsky. However, all those mentioned deserve, to different extents, ownership of the "Minsky moment".

As can be seen from above, equally entitled to be associated with the "Minsky moment" are also Fisher and, in particular, Keynes. Minsky does not bring about significant improvement compared to Fisher in as far as the system of accumulating debt and the inherent endogenous causes of financial fragility are concerned. The Ponzi financing scheme, when short-term debt expands from the need to pay long-term debt, outstanding due to insufficient revenue is found in both. The same is true for the issue of destabilizing calm, the fact that a period of economic tranquillity is suspicious; it prepares, like a trampoline, the conditions for the transformation of a robust financial system into a fragile one. With Keynes, the recognized "Titan", the connection is more complicated.

In his desire to fit into the Keynesian tradition, Minsky lashes the monetarism of the Chicago School. He disapproves of the idea of controlling money supply. The monetarist experiment is found guilty of having eroded "regulation" by favouring "markets". This is how, he believes, the government had to relax the regulations imposed on banks precisely in order for them to acquire the necessary additional compensatory force. And, once liberalized, banks specialized in financial innovations leading to failure. In order to avoid crises, strong control and supervision institutions are needed. He is thinking about the government and the central bank—a saviour and a lender of last resort. "[T]o take care of details"! (Minsky, 2008, p. 117) is the only responsibility reserved to the markets. A fine neo-Keynesian, Minsky sees the central bank as a necessity. It remains to tell us that the bank, just like employment, is a "social asset". But he finds himself, like all those who disapprove of the market and its operating instrument—bankruptcy, at war with logic. He does not like "huge centers of private power" (Minsky, 2008, p. 9); he aims for a game between players of an equal, easy-to-manage size. He describes impeccably how the players, especially the banks, break the rules faced with waves of withdrawal demands. And what does Minsky say then? He argues that the process of refinancing the position (to enable a bank or a financial market to withstand a massive wave of withdrawals) is the essential function of the lender of last resort. After all, it is known that by doing so, the central bank does not eradicate evil, it perpetuates it. Because evil is embodied by those banks besieged by withdrawals. If they had behaved normally in relation to accepted practices, they would not have lost confidence and would not have been assaulted by depositors. Minsky proves he knows two things well: (1) He knows that he cannot rely on the good faith of commercial banks; (2) He also knows that banking innovation goes beyond regulatory attempts. Based on these two correct findings, he suggests two solutions most unlikely to lead to indisputable results: (1) He suggests additional regulations; (2) Even when he finds that a commercial bank is wrong and deserves to be sanctioned, he calls for the help of the central bank to replenish its position. So what is he preparing us for? If you do not eradicate evil, is there any chance of recovery and can we dare to think about sustainability and resilience? No! And Minsky says it bluntly: "[e]very time the Federal Reserve protects a financial instrument it legitimizes the use of this instrument to finance activity. This means that not only does Federal Reserve action abort an incipient crisis, but it sets the stage for a resumption in the process of increasing indebtedness-and makes possible the introduction of new instruments" (Minsky, 2008, p. 106).

At the end of a similar analysis, Rothbard anathematized the central bank. Keynes called for state intervention and monetary injection for the well-known purposes. But his *General Theory* is not a plea, not even an implicit one, for saving, at any cost, from bankruptcy. Therefore, arguing with Friedman, Minsky wants to stay in Keynes's paradigm but proves that he moves away from him as well. Keynes repudiated neither the market nor bankruptcy. For the moment that "produced him", he saw the need for the intervention of the central institutions—state and bank. But, for Marshall's former student, once the storm passed his home remained that of "the liberals and the educated bourgeoisie".

Minsky understands, just like the author of the *General Theory*, that there is no economic peace without social peace; that employment, as such, is not only an economic asset but also a social one. He thinks, similarly to his predecessor, that the state can play a significant role here. Nevertheless, while for Keynes, employment is a function of the actual global demand (investment + consumption), Minsky believes it is possible for the state to create jobs *directly*. As the employer of last resort, the state can also be the depositor of an infinitely elastic job offer. Eliminating the corporate income taxes and the payroll tax and extending the active period for retirees seem to be miraculous solutions. Otherwise, only if Minsky took Keynes's metaphorical solution seriously—bury and dig up the national treasury to create employment at any cost—can we really see in his state a generous bidder, up to a level of full employment! Because, otherwise, it is difficult to detect and understand what are the unbreakable secrets of these jobs created "directly". He does not really relish the idea of "transfer payments" with a cyclical impact on very high government spending. They do not lead to any "direct effect on employment and output" (Minsky, 2008, p. 25).

So where is the solution? The solution resides in a perverted Keynesian mechanism. Similarly to Keynes, Minsky sees in the budget deficit a chance and a source of help for companies and the population in bearing their own debts. The Keynesian multiplier effect is at hand. He sees, like

Keynes, in blatant income inequality a danger to social peace. Just like the "Titan", he calls in the "masters"—the state and the central bank—to lend a helping hand because the market "wiggles" without being able to calm things down and since, clearly, the concern for depositors requires it. The intervention of the central bank holds the assurance that "the losses incurred by a bank or other institution when its assets fall in market value will not be passed through to the depositors at the bank" (Minsky, 2008, p. 49). Interestingly, in line with Keynes, Minsky believes that "[t]he determination of employment, wages, and prices starts with the profit calculations of businessmen and bankers" (Minsky, 2008, p. 285). That is it! He is afraid to involve the central bank. While the *General Theory* paints a mechanism for job creation by simultaneously stimulating investment and consumption as a result of a game between the loan rate and the expected rate of profit, that is the natural interest rate, Minsky leaves the master's line of argument and remains focused primarily on consumption. Why? Because, starting from the results of the practical application of the Keynesian recipe, he gets the impression that "[t]he key to successful policy to constrain inflation lies in knowing that the output of consumer goods is deflationary, whereas investment and government spending are inflationary" (Minsky, 2008, p. 300). We thus learn that the Keynesian stimulation of investments is, by nature, inflationary! Consequently, Minsky is thinking of another mechanism. Taking as key assessment elements the price of capital assets and the price of investment, he believes that "rising short- and long-term interest rates have opposite effects on the demand price for capital assets and the supply price of investment. The demand price for capital assets falls as long-term interest rates increase, and the supply price of investment output rises as short-term interest rates rise. This tends to lower the price gap that induces investment demand. If the rise in interest rates is extreme, the present value of the investment good as a capital asset can fall below the supply price of the investment good as current output. Such a present value reversal, if it occurs, will bring investment activity to a halt. If the interest rate increases are sharp and are accompanied by declining estimates of the profitability of projects, even investment projects under way will be abandoned" (Minsky, 2008, p. 218).

What do we notice compared to the Keynesian mechanism? We notice that Minsky changes the main goal. Inflation, not employment, is of primary concern to him. He does not like the idea of relying on any type of investment because it leads, through an inflationary boom, to a satisfactory state of temporary employment. It is the state of calm in which, "in a capitalist economy that is hospitable to financial innovations" (Minsky, 2008, p. 199) endogenous forces appear and break the peace; they turn a robust economy (one in which income flows predominate to meet commitments) into a fragile economy (one in which portfolio transactions "create" balance-sheet means of payment). It is not the open market and the minimum state, plus the general reduction in profit tax—Laffer's legacy—that constitute his weapons against Keynesian inflation. No, his solution is increased consumption and low investment! Because, he notes, "inflation can be slowed or stopped by increased production of consumer goods" (Minsky, 2008, p. 288). He is not indifferent to investing in productive assets. He believes and writes that "[a] main characteristic of a capitalist economy that is stagnant and or immersed in a deep depression is that the 'capital development of the economy' is not going forward" (Minsky, 1992, p. 15). Delaying the growth of productive capital is done in favour of access to and ascendancy of speculative capital. These high and immediate profits dislocate significant funds and credit flows from the real economy to the acquisition of financial assets. Phenomenon that largely leads to financial instability, the basic Minskyan thesis. That being the case, if we have problems with investment, he believes, we focus on consumption.

Who should consume? Everyone! Along with bank supervision by means of the "discount window", supporting competition between small and medium-sized banks, encouraging financing through the issuing of participatory securities as elements of his work agenda, Minsky is concerned with the situation of those with low wages and child benefits. In order for incomes to reach those who consume, he believes it would be appropriate that the growth rate of high wages should be below that of labour productivity, and in the case of low-wage workers, wage increases should exceed productivity gains. Socialism! Not scientific, but utopian.

Socialist temptations are a personal issue of the one who set out to show us the way to stabilize an unstable economy. We are concerned with the novelties Minsky puts forward in support of sustainability. However, if rejecting Friedman, ignoring all the classicists and all the "Austrians", speculating between Fisher and Keynes, invoking in support of his argument outstanding names such as Alvin Hansen, Lawrence Klein, John Hicks, Paul Samuelson, Don Patinkin, Franco Modigliani, James Tobin, Michael Kalecki and others, and in the end he suggests growing mostly through consumption, without taking into account productivity in stimulating those who produce, we believe we need to reflect more on the "Minsky moment" when we do our homework on sustainability.

2.3 Can We Find Sustainable Ideas at the 2008 Crisis Theorists?

Certainly, we are interested in what the great voices of this period are saying. Similarly to the 29–33s, when the last great rupture in the world economic dynamics took place, there was a storm of ideas, on a suspiciously varied range and playing upon the same object—the crisis. What "news" bearing a message towards sustainability do we find at the new crisis theorists produced by "The Great Moderation"?

First, we are revealed new facets on the issue of indebtedness. *One* refers to Bernanke and Gertler (1995). They are "rediscovered" by neo-Keynesians, to show that credit does not create debt according to the Minsky scheme, but as a financial accelerator through interest rates. Moreover, not as an independent phenomenon, credit enters the machinery of banking mechanisms for transmitting monetary policy in multiple ways: the bank credit channel, the bank balance-sheet channel, the non-bank balance-sheet channel (Boivin et al., 2010). *Another* facet concerns the distribution of debt. In the case of a closed economy, the total debt, in accounting terms, can only be zero because the debt of one economic player is another one's receivable; a debt to themselves. However, Eggertsson and Krugman (2012) point out that indebtedness is not uniform, on the contrary. And, in this context, only the unequal distribution of debt can make some debtors unable to pay their debt from proceeds. That is, not all borrowers enter the same crisis-generating indebtedness. Perfectly true!

Schumpeter, a strong supporter of the idea that credit is an important stimulus for growth, is also rediscovered. Rediscovered by the hypothesis that a loan, serving the innovative *entrepreneurial* process, *creates* purchasing power; it is not just a *transfer* of purchasing power from the savings liabilities. That is, Hagemann (2013) tells us, we are dealing with a process of creating purchasing power "out of nothing". If credit money turns into innovation, which, in turn, renders an economy more dynamic—on its real and not speculative side—the process remains healthy; we are no longer dealing with a Ponzi scheme! "It has no room for speculation", Minsky argued precisely in this regard

(Minsky, 1992, p. 18). Linking Schumpeter's name to Minsky's, Steve Keen (2009) believes in a "Schumpeter-Minsky law" according to which loan-generated aggregate demand is a sum of pre-lending income and newly employed credit flow. The lower the share of credit demand and the higher the share of previous savings of the total demand, the more likely it is for a trend towards sustainable development to occur. This idea is to be found, in other words but on the same logical structure, in the works of Mises and other Austrians.

Even if Steve Keen does not bring new information on the nature of the aggregate demand surplus generated by lending, he still deserves attention where he explains the unsustainability of real estate investment lending. What is missing here, Keen believes, is a liquid competitive market. Prices are formed conjecturally; they do not follow the logic dictated by the supply–demand tension. What is missing here is the *simultaneity* of a large offer that would lead to price fall. Properties are offered one at a time and chaotically. The same is true for the proceeds from such business. The outcome? Too often, the ability to repay old loans becomes a problem, and that is how we get into recession! (Keen, 2007).

Mishkin (1999) completes the picture, warning that the well-known phenomenon of adverse selection and moral hazard, a result of asymmetric information between creditors and debtors, exacerbates credit risk during periods of financial instability. In addition, it tells us a little about the dangers of the "liquidity trap". The idea of giving up by some central banks to claim minimum required reserves on the grounds of the inability to influence the credit offer developed by another Minskian supporter, Wray (2001), is also worth all the interest.

What other authorized and prominent voices say is no less important. In *The Return of Depression Economics and the Crisis of 2008*, Krugman (2009) attempts at the liberal origins, or what's left of it from Keynes, accepting the "fine-tuning" that the jammed economic engine could be fixed by "a very limited sort of intervention—intervention that would leave private property and private decision making intact" (Krugman, 2009, pp. 101–102). How much is left of private law in deciding, in a "Keynesian pact", is known. In line with the same mentor, Keynes, does not disapprove of the idea of the saving press, if need requires (Krugman, 2008). Comparing the market with state intervention, he believes that the market must be allowed to solve problems before panic sets in. If the fire escalates, the market no longer helps and intervention is needed (Krugman, 2007). Faced with the "unleashing hell", another

Nobel Prize winner, Nouriel Roubini, is just as staggered. The prophet putting forward a pioneering work in announcing the crisis is animated, to a certain extent, by the spirit of the free market. With assertions of the kind "it is absolutely necessary for insolvent banks, firms, and households to go bankrupt" (Roubini & Mihm, 2010, p. 31). The authors seem to believe in the healing virtues of bankruptcy, an essential attribute of the free market. They even seem outraged by the claims of the aided banks, saying that "[i]t's like putting lipstick on a pig" (Roubini & Mihm, 2010, p. 89). However, worried that these "zombies dependent on public credit" (Roubini & Mihm, 2010, p. 89) will pull all the fuss, and "burn" the whole economy, Roubini admits, just like Krugman, the intervention of "firefighters"! Compared to them, Stiglitz is more incisive. The endemic fundamentalism of the market bothers him. Deregulation, beginning with the Washington consensus and continuing with the Phil Gramm-Leach-Bliley Act, with all the underlying derivatives, one more toxic than the others, turns him into a harsh critic of neoliberalism. "Socialism cares about people (…) 'corporate welfare-ism' (…) it's trying to help corporations, not people" (Stiglitz, 2008). We, authors of these lines, know well how good socialism (in its most "noble" version—communism) has been for people. But, we admit that in distress, while the fire of crisis is burning the economy down and gives people a lot of grief, you can also think of such an alternative; one in which the fire has nothing left to burn!

2.4 Concluding Remarks

Therefore!

The founders' lesson about the sources of economic dynamics and its self-propelled forces is sound. The economic civilization has found nourishment in it. The crisis of 29–33s seriously challenged its claims and hypotheses. The dispute to find its causes at the time is well known. Too much or too little market? To be, or not to be! Allowing theorists to spend a lot of words to impose their own opinion, reality has imposed its own lesson. And the lesson for the beginnings of twentieth-century capitalism had a new name: Keynesianism. Unjustly assimilated to pure statism and total interventionism, Keynesianism replaced the old and verified founding lesson. The neoclassical synthesis works produced under the umbrella of P. A. Samuelson did not dispel the feeling that a great lesson had been lost. There is yet another interesting aspect though. There are many areas where probing is engaged for the project of a generous lesson on sustainability and resilience. Those who animate these searches, with dim exceptions, do not go back. Smith, Ricardo, Say, Malthus, Bastiat, etc., they all seem lost. Keynes remains running with all this. Could the fact that he is closer to the classicists than the so-called neoclassicists—as Mason and Butos argue—explain this propensity for the author of the *General Theory*? Or does the cumulative process, the fact that the strong ideas of those who laid the foundations of economic science are found in the present theoretical bodies, spare us to turn to the "archives"? Maybe both! To find an answer to these questions, we selected a few moments from the contemporary history of economic thought.

We stopped, first, on the *heavy* topic of the economic cycle. Its status as a subject belonging to the hard core requires an equally worthy approach. It requires a Cartesian, geometric spirit, clearly defined causal relations, similar to those in the area of heavy sciences. We were puzzled to find out that this was not the case. Instead of dry rationalities, we found a linguistic Babylon. We found too many places where interested philosophies confiscate or subjugate analyses, depriving them of the possibility to send to us indisputable essences and articulate conclusions under the matrix of sustainability.

For example, the new Austrians seem more interested in rebelling against their own spiritual parents than in updating and building on their inherited dowry. They forget almost completely about the natural rate of interest and its fantastic role in shaping healthy economic dynamics. And this, in order to specialize. To accurately present the inherently unsustainable mechanism of stock market-like capitalism based on pyramid games. Nowhere do we find a more vivid portrait of the fraudulent entrepreneur who supports a parasitic mechanism called the Central Bank. The idyllic path of a gold or silver currency cycle, as a premise, is also a work of art. However, these specialized parts do not coagulate. Austrian contemporaries quarrel with each other; they invoke ideological enemies in defence; they fight over Friedman's arguments and see visible and invisible "errors" everywhere. They only resonate with a certain position when they remember that these occur in the shadow of the master: money does not measure anything; it only acts as an intermediary—Menger whispers to them. And to prove obedience, they obsessively recall his arguments and tell us, also obsessively, that "any amount of goods can be changed with the help of any amount of money". The entire amount of goods in the USA can be exchanged for only $100! If anyone thinks that such sandy formulae can be the solid grounds for sustainability, we argue that it is an unfortunate illusion.

We found Milton Friedman more confident in the chances to gain stability. His methodological inventions support him in this direction. Without naming his attempt (at least the name of Ricardo should have been mentioned), he tries to revive the fundamental classical relationship between the volume of production and the amount of money. Although he denies, at least at formal level, his paradigm interferes with Keynes in the area of anticipation. Fisher is no stranger to him either. The analysis also reveals a lot about Wicksell. But unlike the Swedish economist, he does not have high hopes for the leverage called rate of interest. He sees in it a deceptive tool. He prefers the money supply as an alternative. But even here, prudence and not discretion shows the extent. He also remains interesting when, by reinterpreting and adopting the Wicksellian concept of the natural rate, he puts forward a credible analysis of the relationship between inflation and unemployment. Overall, we find him accessible and useful in finding a new lesson about the healthy dynamics of the economy. Especially in places where, obviously, it is clear that he is not willing to sacrifice the real for the nominal. And his real aims at employment. With unemployment in the yard, Friedman's economy cannot claim sustainability.

We can find exploitable areas in the works of Fisher as well. We refer, first of all, to all the points of interference with Friedman; not so much to his famous equations, as to his concern to link the level of prices to the quantity of money. The personal note in which he approaches the social dynamics is equally worth mentioning. The idea that once caught in the dynamics through integration and globalization as a poor man, you are stuck this way, invites reflection. It is worth looking for the reasons why in the new era of globalization and knowledge, Fisher believes that the social dynamics based on Adam Smith's scheme remains a mere aspiration. The analysis of indebtedness and unindebtedness, on how over-indebtedness followed by deflation outlines the evolution axis through the cycle renders Fisher trustworthy. And this, in a world where the bank is hungrier and greedier and a more skilled loanshark.

Under the heading "renowned interpreters in search for sustainability" we stopped on the works of Mason and Butos to show how if we get caught up in doctrinal quarrels, instead of passing on well-articulated theoretical structures to future generations we only serve them salads of ideas. Because this is what we do when, suffocated by doctrinal passions, we rebelliously oppose the "rational absurdities" of Friedman's monetarism to the undeniable clairvoyance of neo-Keynesianism. Some have nothing but flaws, others know everything!

The Minsky moment struck us through its seductive record in terms of citations that diverge with his disarming project to find solutions for a healthy policy. *Stabilizing an Unstable Economy* is a grandiose project. The way in which speculative financing is a cause for the weakening of the financial system is grasped by Minsky in an excellent analysis. However, if the picture deserves this label, the film, and, in particular, its ending, is a big disappointment. Deliberately placed in the shadow of "titan" Keynes, Minsky sends us to a place where not even the master had been. To fight all evil in the world, the author of *Stabilizing an Unstable Economy* suggests measures that go beyond the boundaries set by the *General Theory*. Relying on heavy regulation and a saving bank of last resort, Minsky is still in the titan's yard. With jobs created "directly" by the state, where consumption is used to achieve growth, with the anathema thrown on a market that is too "wiggly" and with a suspicious way of correlating productivity and wages, Minsky becomes original; he moves away from the master and, to the same extent, from the requirements of a lesson pretending to deliver a credible message.

We have noted from the acknowledged proponents of the last great crisis that not even in the harshest moments of the economy are we able to remember the founders and their splendid lesson. We revive the works of Keynes or Schumpeter. No word of Smith, Ricardo, Say, Marshall; by no means the "inedible" Wicksell. All in all, crisism is a firefight attempt by which Keynes's state is called to put down the fire. The machine gunning of the market fundamentalism is a key ingredient fuelling fire extinguishers. The doctrinal dream, taken from Stiglitz, of a socialism that takes care of people can only serve as setting! What about sustainability and resilience? They do not enter their scene!

References

Bernanke, B. S., & Gertler, M. (1995). Inside the black box: The credit channel of monetary policy transmission. *Journal of Economic Perspectives, 9*(4), 27–48.

Boivin, J., Kiley, M. T., & Mishkin, F. S. (2010). How has the monetary transmission mechanism evolved over time? In *Handbook of monetary economics* (Vol. 3, pp. 369–422). Elsevier.

de Soto, J. S. (2006). *Money, bank credit, and economic cycles*. Ludwig von Mises Institute.
Eggertsson, G. B., & Krugman, P. (2012). Debt, deleveraging, and the liquidity trap: A Fisher-Minsky-Koo approach. *The Quarterly Journal of Economics, 127*(3), 1469–1513.
Fisher, I. (1930). *The theory of interest: As determined by impatience to spend income and opportunity to invest it*. The Macmillan Company.
Fisher, I. (1933). The debt-deflation theory of great depressions. *Econometrica: Journal of the Econometric Society, 1*, 337–357.
Friedman, M. (1968). The role of monetary policy. *American Economic Review, 58*, 1–17.
Friedman, M. (1970). *The counter-revolution in monetary theory*. Institute of Economic Affairs.
Friedman, M. (1993). The "plucking model" of business fluctuations revisited. *Economic Inquiry, 31*, 171–177.
Friedman, M., & Schwartz, A. (1963). *A monetary history of the United States*. University of Chicago Press.
Hagemann, H. (2013, March 13). *Schumpeter's theory of economic development*. Paper presented at the University of Kragujevak.
Hayek, F. A. (1967). *Prices and production*. Augustus M. Kelley Publishers.
Hülsmann, J. G. (2008). *The ethics of money production*. Ludwig von Mises Institute.
Keen, S. (2007). *Deeper in debt. Australia's addiction to borrowed money*. The Centre for Policy Development Occasional Papers (3).
Keen, S. (2009). Policy forum: Household debt: The final stage in an artificially extended Ponzi Bubble. *The Australian Economic Review, 42*(3), 347–357.
Keynes, J. M. (2018). *The general theory of employment, interest, and money*. Palgrave Macmillan.
Krugman, P. R. (2007). *The conscience of a liberal*. W. W. Norton.
Krugman, P. R. (2008, March 14). Betting the bank. *The New York Times*. http://www.nytimes.com/2008/03/14/opinion/14krugman.html?_r=1
Krugman, P. R. (2009). *The return of depression economics and the crisis of 2008*. W. W. Norton.
Mason, W. E., & Butos, W. N. (1996). *Classical versus neoclassical monetary theories: The roots, ruts, and resilience of monetarism—And Keynesianism*. Springer Science & Business Media.
Minsky, H. P. (1992). *The capital development of the economy and the structure of financial institutions*. Hyman P. Minsky Archive, paper 179, The Jerome Levy Economics Institute of Bord College.
Minsky, H. P. (2008). *Stabilizing an unstable economy*. McGraw-Hill.
Mishkin, F. S. (1999). Global financial instability: Framework, events, issues. *Journal of Economic Perspectives, 13*(4), 3–20.

Rothbard, M. N. (1990). *What has government done to our money?* Ludwig von Mises Institute.

Rothbard, M. N. (1994). *The case against the fed*. Ludwig von Mises Institute.

Roubini, N., & Mihm, S. (2010). *Crisis economics: A crash course in the future of finance*. Penguin Press.

Stiglitz, J. (2008, November 5). Crisis points to need for new global currency, interview with Kiyoshi Okonogi. *The Asahi Shimbun/The International Herald Tribune*. http://www.asahi.com/english/Heraldasahi/TKY200811050060.html

Wray, L. R. (2001). *The endogenous money approach*. https://ssrn.com/abstract=1010328 or https://doi.org/10.2139/ssrn.1010328

CHAPTER 3

How to Conceive the Brundtland Agenda in the Context of the Nominal Economy's Imperialism

This journey carried out on the theoretical framework of those who seriously pointed out the evolution of the theory of economic dynamics had a purpose. We wanted to find out what was the end of the road for the vigorous ideas that, inspired by the ideational core of the great founders, were just waiting for some polishing and updating. And what did we find? We found ideas or large fields of ideas abandoned; or truncated; or mixed with ingredients until no trace was left of them in the final composition. Or, leaving out any minimal alignment claimed when it comes to leading issues, it is visible from miles away that only the Tower of Babel can compete with them. It is a known fact that the desire to provide a lesson about a good understanding of the economic world is related to the ferment that keeps the economic science alive. But with a devastatingly diverse lesson in conflicting opinions on the same subject, its applicability becomes questionable. We rely on the founders in order to find energy sources and the chance to make it attractive to a practice that today seems to be healthy precisely in places where it does not take into account its statements.

In all its sequences, the lost lesson would receive robustness and a chance to vigour through this comeback. An attempt to reconfigure the Brundtland agenda and translate it into new realities will undoubtedly

I. Pohoaţă et al., *The Sustainable Development Theory: A Critical Approach, Volume 2*, Palgrave Studies in Sustainability, Environment and Macroeconomics, https://doi.org/10.1007/978-3-030-61322-8_3

argue this assertion. An allowable prioritization will lead us to three sensitive areas with great potential in shaping a trend towards sustainability and resilience. Among these, the natural rate of interest has, by far, priority.

3.1 How Would an Economic and Sustainable Up-to-Date Agenda Look like If Conceived According to the Bruntland Matrix?

Drawing on works belonging to different or even opposite theoretical-doctrinal spaces, we believe that an economic agenda burgeoning with ideas that could be reflected in the Brundtland agenda might look as follows:

- Free, competitive market, with competition between actors that: (1) gives meaning and content to the process of price formation according to the guide of the natural price of goods and money; (2) prevents the occurrence of the "too big to fail" situation. Monopoly is inevitable in a free market, according to Schumpeter's view of innovation. This purpose is not problematic if the "giant" is allowed to go bankrupt when it does not comply with the laws of the market; (3) puts the institution of bankruptcy back in operation;
- Small but strong state, responsibly positioning the economy within firm and rational rules, as well as clearly defined property rights; a caring state, defender of free competition and an enemy of monopoly;
- Accepting the idea that the state can create jobs "directly" only in fields like security, protection and other public areas par excellence;
- Reconciling modern economic theory with the theory of money as a standard of value and discrediting, in this way, the "rationality" of the discretionary monetary policy and the printing press as a means of saving those who do not deserve to be saved;
- Recompacting the fiscal policy with the monetary one;
- Reduction of taxation following the argumentation that imposes this idea: money sources intended for state taxation can be more efficiently used by taxpayers—except for the strictly public goods only;
- Promoting and encouraging the saving spirit in order to acquire the capacity for development drawing on voluntary savings;

- Promoting the 100% compulsory reserves policy when loans above voluntary savings become necessary;
- Unlocking the "mystery" of actual demand, sending money to those who require it for productive purposes, engaging in policies that stimulate both demand and supply, eliminating, on this route, the alleged tidal or "helicopter" effects in price formation;
- Repositioning the entrepreneur as bearer of the first impulse, a landmark and a starting point in any calculation about the natural price and the natural rate of interest;
- Resettling the central bank, institutionally, in the position of a tool in the service of the economy and not of an entity in itself; competition between small and medium-sized banks;
- Transforming the natural rate of interest into a priority guide for the Central Bank;
- Promoting the idea of macro-equilibrium as a result of sectoral equilibria: savings vs. investments; capital goods vs. consumer goods; revenues vs. budget expenditures; price level vs. production level;
- Imposing the premise that technical progress: (1) allows a healthy cut in prices through low costs for high productivity; (2) changes the production structures, the relationships between factors and the managerial philosophy;
- Employment is a heavy part in the artillery of an economic policy; it must remain a resultant and not a starting point in a business;
- The monetary impulse is important but inflation has sources along the entire economic process; it is important to know where it proliferates;
- Money out of nothing certainly leads to inflation and crisis and the budget deficit does not foster sustainability;
- The crisis is not a fatality! The founders provided arguments in support of this idea.
- Etc.

3.2 Three Delicate Areas Where the Recourse to the Founders Would Lead the Way Towards Sustainability and Resilience

An epistemological conflict requires resolution either by appealing to a model or a school, or both. In the ideational chaos we tried to set order to, we cannot conclude that the ideas that are likely to support sustainability prevail over the others. But we can say that they are, most of the time, dissipated and centrifuged in such a way that, from this dilemmatic scattering, only a master metaphysicist can turn them into a coherent and credible scientific product. Is such an approach predictable? Since the present is too full of experts at complicating things, the first answer would be NO. We believe that a positive answer is possible as well. And we also believe that not much needs to be done. It should be noted that old truths were obscured by the passage of time, although they could have served as landmarks. In order to discover the right path to sustainability and resilience, we need to rediscover them, by going back to the founders. What are the areas that need such action most? At least three, we would argue: (a) the role of money; (b) the role of the great players on the economic scene; (c) the beacon role fulfilled by the natural rate of interest. We believe that in these areas, validation by updating the works of the founders is more necessary and more stringent.

3.2.1 Playing with Money When Dealing with Inflation and Distribution

The "call of the ancestors" is more urgent here than anywhere else. We have shown in the previous chapters that in the classical line of thought, the function of money as a standard of value was part and parcel of the very logic of money origin. The other functions, including that of means of exchange, would further derive from it. Using this hierarchical structure, with the standard of value at the forefront, the classicists provided sustainable analyses. They also got confused by it—Ricardo is clearly a case in point—when, with the help of this function, they attempted at an absolute determination of the value measure. Giving up the objective value and moving on to the subjective value was a major accomplishment on the neoclassical agenda. However, placing the medium of exchange function on the top position and obsessively giving up the standard of value—with a load of arguments—did not open the doors to sustainable analyses. On the contrary, it provided support for confusing, argumentative, often contradictory and implausible positions. The paradigm shifts and the attempts to give primacy to the means of exchange function belong to reputable names or schools, which occasioned a dangerous game with money. Not only on a doctrinaire pathway, but also beyond it, the function of money as a means of exchange became the working tool of all neoclassicists and neo-neoclassicists. Those who define, today, the mainstream of economic science relies on it as well. Marx's observation that for the neoclassical school the meaning of money remained unclear is groundless. No, the neoclassical theory of money was and remains clear; it is the one built on Menger's ideational framework (Mason & Butos, 1996).

It would be nothing serious if such a function supported both coherent theoretical registers and sustainable policies. But it does not! The broken logic of Menger's early beginnings constitutes the support of inconsistencies, inaccuracies and strange hypotheses that, all together, scientifically regulate a game with known ending: the evil produced by the game with money out of nothing! In addition, it is precisely such an argumentation that entangles things on two important levels of economic policy: inflation and distribution.

Insofar as money as a means of exchange does not measure anything, what is the definition of inflation then? If there is one in the first place! The perception—a textbook type—of the phenomenon translates into a generalized, permanent and unacceptable increase in prices, within the limits of a "normality" of up to 4–5%. Perhaps we anticipate and find such a meaning for Friedman's inflation. He is afraid of inflation. His end of journey with "expensive money" confirms his fears. Reality also confirmed to him the fact that we are fighting today with inflation triggered yesterday. Fisher thought that deflation was more dangerous than inflation. Laffer also struggled with soaring prices. Up to the point where the extra money reduced the rate of interest below the rate of profit, for Keynes inflation was something that "didn't hurt". What does Minsky mean by inflation, that he suggests to reduce it by redistributing income and increasing consumption? It is far from easy to understand. The biggest confusion is in the works of the Austrians. If neither the monetary volume nor the general level of prices matter, then what does inflation mean for them? Hülsmann argues that it is an increase in the amount of money beyond what the free market establishes. Besides, the free market also determines the optimal amount of money. And the optimum is, if we are to believe him, any amount of money. You can have inflation with 100 units and no inflation with two hundred billion units, if these two hundred do not exceed the limit set by the market. Rothbard puts forward a more "neutral" attempt. First, he admits that it is acceptable and beneficial to reduce prices by reducing costs. Then, for him, inflation is synonymous with counterfeiting, that is, creating new money out of nothing; new money that flows into the economy, first to counterfeiters and then to those who pay tribute to the phenomenon. Whoever gets the new money first earns through inflation. Those at the end of the queue are penalized. At the same time, inflation is functionally synonymous with taxation, an act of dispossession, to which governments resort directly, annoyingly and electorally as a much more refined despoilment. All in all, the meaning of paper money is a contract with a declared inflation and an end of the road in "the order of a cemetery" Hülsmann (2008, p. 172) believes. As a result, getting out of inflation means gold, not paper. And the prospect of a return to gold indirectly implies a return to money as a standard of value.

Would we have so many definitions of inflation if money were the standard of value? No, and this is not even possible.

The way in which the money river flows, fuller or drier, faster or slower, is also related to the distribution. Why? Because what is distributed, no matter in how many stages, is the money! Goods and services come after; actual goods are purchased with the money received. And if we do not operate with a standard of value, what do we distribute? We distribute a means of exchange, whose purchasing power is either the same or does not matter at all? Moreover, if the total volume of means of exchange is not related to the volume and value of the goods, expressed by prices, then based on which (supposedly rational) criteria is the distribution directed? It is clear that it is turning into a difficult problem, and perhaps even the most important one, as Ricardo thought. The classicist looked at it from the perspective of the conflicting character to which the objective theory of value relegated the problem. We have to address it from the perspective of the manoeuvrable character thanks to which distribution, via inflation, can deceive. It can create false impressions about the generosity of some governments. It can create the illusion of growth by multiplying goods of false value, expressed in prices by money that measures, as a means of exchange, the value of wind as well as that of bread. Nothing is more unsustainable than the lack of benchmarks.

The "voice of the ancestors" and the orientation towards gold, supported, paradoxically, precisely by those who broke away from the standard of value function, underline the urgent need for benchmarks. Without them we can only deceive ourselves with illusions, blaming, in turn, the market and the government. But we will not get rid of bubbles, money out of nothing and electoral deception if the standard of value of money remains in the dustbin of history.

3.2.2 *Masters of the Economy. From Instrument-Actors to Masters that Set the Rules*

The "main characters" we are talking about are the central bank and the government. Their brief history presents them as tools; human creations designed to serve the individual and the collective welfare. All the efforts of the New Institutional Economy serve this purpose: to show that the two are important players in social and economic life, with a strong institutional load, made by man but also rule-makers, or institution-makers. Hayek's (2012) trilogy, *Law, Legislation and Liberty*, correctly and unquestionably establishes their place in a state governed by the rule of law. In short, the economy and society need the two players. Even the most ardent defenders of individualism and open society accept them. But not under any circumstances, but in conditions specified in the job description, likely to keep them on the leash and make them responsible. Who completes their job description today and to whom they are accountable are very important questions. We are interested in an attempt to answer them insofar as the behaviour and influence of the two major players are related to sustainable development and, especially, to resilience.

What are we really interested in? We are interested in finding out why they emerged, what they are and what is needed for them to be considered pillars of sustainability.

At School we learn that the bank is the good character who replaced the evil loan shark; that everything we found repulsive in the usurious spirit we no longer meet at the bank. That is, we enjoy going to the bank. Why? To make deposits under guaranteed conditions, to borrow in decent conditions and to succeed in a business.

There is another story behind the history of the emergence of the most famous banks in the world. Reasons to buy government debt securities explain the beginnings. Beginnings that, as is shown everywhere, "sentenced" from the very beginning the central bank to a cohabitation with the state, a relationship that the two partners cultivated in the formula of a successful marriage of convenience; firstly, convenience for them. Cantillon and Turgot wrote thoroughly about it. Rothbard's well-known incisive book, with references to the history of the Bank of England and the FED, sets out the main steps towards the most perfidious and solid marriage in the world, as follows:

- The first step, *prenuptial cohabitation*: opening the money tap intended for the purchase of public debt securities as the dependence on voluntary saving was slow and was hindering business;
- Step two, *functional marriage*: the government lends a saving helping hand to the bank, allowing it to suspend payments;
- Step three, *gifts are offered*: the government gives the central bank an exclusive right, monopoly over currency issuing; thus, it becomes not only a distributor of funds but also a money issuer;
- Step four, *the bride's dowry*: the government links banks to the umbilical cord of the Central Bank, forcing them to open current accounts. The inverted pyramid system is set up and the Central Bank becomes a certified lender of last resort. It is given leeway to issue more money, as "needed";
- Step five, *sealed marriage*: the central bank becomes a bank cartel, imposed by the government and desired by bankers; it is liable for the "power of money" and the "lack of liquidity"; it becomes the depositary of state accounts and is authorized for open market operations. Depriving the state from authority and prestige, it also receives from the "groom" the implicit promise that it is far too important to let it collapse!;
- Step six, *marital frictions*: banks complain about too much wellness! They want, from their protector, more involvement in the monetary expansion. Careful with the "power of money" but also with the worthy "daughters", the central bank learns from the government that there are also political pressures. An "injection" and everything ends well. The marriage is unshakable! (Rothbard, 1994).

The ultra-synthetic excerpt from the analysis carried out in the book cited above is of great use to us. It shows us that by maintaining the appearance of ethical responsibility and involvement for the welfare of individuals, even in a consolidated democracy, an insidious relationship can be concluded and maintained between the government and the central bank, contrary to all that the spirit fostering sustainable development means. In other words, the result of the mentioned marriage is wrong: two saviours of last resort of players which, in a free market, must be allowed to compete and remain in the game only if they follow the rules, not saved because they are too big. On the one hand, there is the Large State, concerned with the social welfare to such an extent that it can borrow to pay for electoral handouts, to create jobs "directly" as a first-instance employer, regulator and protector of large corporations—industrial or banking—involved in income distribution and employment policies—including in the private sector—sold to the banking system whose printing press is just as vital as air. On the other hand, the large and strong Central Bank is present as a generous creditor, and state depositary; supervising the movement of prices and everything possible; a centre of power, diffuse but present, around which all the economic players revolve.

The shared fashionable idea is that a strong economy is one with a strong banking system. There is no denying this. As long as the bank is necessary, it is admissible to believe that it should be strong as well. This is not the problem. The problem is that the perfidious nature of its relationship with the government defies expectations and is contrary to the resilience of the economy. The bank is too strong in relation to an enslaved government. By diligently cultivating its interests, the central bank came to validate Ricardo's prophecy. It does so as an institution that does not account to anyone *de facto;* it has its own budget from which it pays its employees according to its own criteria; its board members are elected professionally, by competition, but also on political "grounds" with interests in the area of parties, government, presidency or the large corporations; a micro-level oasis of a godfathership capitalism, in which all those interested contribute to completing the job description. From the position of saviour of everything, the Central Bank acquires an aura; enough as to convince the public that it is normal for it not to account to anyone; not to receive advice that would "weaken" its independence. Technical independence, YES, but it wants more than that. To be strong in the fight against the monster—inflation, it must not be encumbered! Neither when it finances deficits, the electoral cycle, nor when it encourages the formation of cartels. In order for this to be possible, "doctrinal perverting" is required. Reputed scholars need to draw and support its status as a fighter with scientifically proven chances of winning against the imperfections and "waste" of free markets. And many are those interested in signing the payroll offered generously, and without mimicking common sense in the conditions in which even the marriage of the bank with the Throne or Altar played well with the public. No matter how many arrows come from those who try to reposition it, to place it in the good and necessary classical tradition, the results are feeble.

After all, why would the fact that the Central Bank and the government have become the masters of money upset the defenders of sustainability and resilience? There is no reason why, that should be the answer. But it is disturbing because something bad came out of that marriage. From the status of mere instruments, the bank and the government have become absolute masters of money and not only. And, in this capacity, they promote and support inaccuracies, rules inconsistent with the spirit of the free market, misleading paradigms that are contrary to sustainability. *Here are just two hypotheses*:

First, there is the perfidious cohabitation between the Central Bank and the Government which distorts and undermines the legislative process. What reactions and legislative initiatives should we expect noticing that the bank collects interest for money it doesn't own or it manipulates counterfeits by turning them into credible sources of financing? We should expect *Laws*, rules. Laws as imagined by Friedman and Phelps that would not allow an uncorrelated increase in prices in relation to production; clear provisions on the permitted conditions for bank cartelization; laws sanctioning false certificates and receipts, fractional reserves, recording movements of false values in the balance sheets, and so on. Do we have something like that? No! No, because this is the area where law, fraud and reputation are caught in a causality that is just as treacherous as the game in which it occurs. Here the movement of interests dictates the change of the rule, not the other way around. And interests are combined. The bank needs a reputation before it dares to play fraudulently with money. Reputation comes from its behaviour, but it may also be "allowed by the master". The master, that is, the government, promotes it because it is interested in getting more money whenever it wants. This means that law-making is full of influences. And for the reasons mentioned above, it is missing when it should be present. It is not difficult to regulate credit expansion in the territory, the massive production of toxic bubbles and insurance based on nothingness or the fractional reserve and cartelization. But the Central Bank does not propose it and the government does not do it because it would also lose definitively. And without rules, without good practices, sanctioning every outgrowth on the economic body, sustainability and resilience remain in the dream drawer.

Second, the way in which the government cohabits with the Central Bank and the banking system in general explains, as a matter of priority, the *bankruptcy of the institution of bankruptcy*. Nothing more dangerous for a system we wish it were healthy. The analysis of the process by which the Central Bank became one of the first decision makers without following any orders shows that not only is it sure that it will never go bankrupt—its bankruptcy would be equivalent, within this paradigm, to state bankruptcy—but that it has the baton and the willingness to determine who and how other players deserve to be treated when faced with bankruptcy. And it makes sure that those who take advantage of it and return the favour do not go bankrupt. There is no need to argue upon the importance of rehabilitating an economy and its chances of recovery through the bankruptcy of those who broke the rules of the game and used resources unsustainably. It is like thrusting the surgeon out when a trepanation or scalpel extraction of an inflamed appendix is imperative. However, deregulation and sending companies, including banks, to the competitive landscape and, naturally remedying and correcting deviations, have been qualified as fundamentalist! In such a way that it seems inappropriate and senseless to talk about the fundamentalism of the government or the Central Bank. On the contrary, through the contribution of some Nobel Prize winners in economics we can find out that rationality can only stay in their court. For example, between individuals' rationality and the rationality of the government, Stiglitz bets on the latter. The behavioural fluctuations of individuals, on a "systematically irrational" background, would even explain the crisis: "irrational exuberance leads to bubbles and booms; irrational pessimism to downturns" (Stiglitz, 2010, p. 110). The Government and the Central Bank, but especially the former, with its tanks full of rationality, can transfer some of its overflow to the disoriented and lost citizens! "Government has an important role to play: it should not only prevent the exploitation of individual irrationalities but also help individuals make better decisions" (Stiglitz, 2010, p. 110). When we conclude that a deficit of individual lucidity is filled with state generosity, in alliance with a Central Bank, then we have a problem. A problem with what the new institutionalists, but not just them, call procedural rationality (rule-following behaviour); a solution to the acknowledged limited rationality of individuals. It is required that, through mimetic behaviour and complying with the rules, individuals with deficits correct their behaviour. But this if, and only if—a great condition which Stiglitz does not speak of—the rule, the practice urges to be followed because it is the embodiment of rationality. But following a rule made by a government resorting to a marriage of convenience with a bank, driven by its own interests, this equals complete perversion. For "state or banking reasons", individuals agree that firefighters put out the fire and the bank or company they work for is rescued.

We admit that there are state reasons, and even banking reasons. But not per se. They are charged with meaning only through and in connection with individuals. Beyond them, rationalities are empty. It is said that the reason for an optimal budget is for the state treasury to be full. If the individuals of that state are doing badly, the "reason" mentioned is based on moon rays! We are talking about the same kind of "reason" when the economy is in "freefall", to use Stiglitz's terms, but nothing happens to the central bank and the government.

No one is questioning the fact that strong banks and credible governments provide the foundation for sustainable development. Without them, it is difficult to imagine how development projects can be promoted. But not just in any way. When they join hands to promote an inflationary culture—a synthesis of this treacherous game in two—the economy winds up on its "immunity" weakened by this diabetes sent upon it by Midas by means of the two ambassadors and it suffers a crisis. All goes well, that is, until from mere instruments the two turn into and acquire the behaviour of masters; they promote their own "reasons" and turn the institution of bankruptcy into a ghost. And an economy without this institution—in a functional state—is like a community without doctors; it runs the risk that, once a poisonous and toxic mushroom appears—a deceptive bubble—it sends the economy, we repeat, in "freefall"!

3.2.3 Achieving Sustainability Through the Natural Rate of Interest—A Necessary Synthesis of Ideas

Nothing can capture, in a more condensed and clearer way than the natural rate of interest, whether or not we are in the field of sustainability. If Ricardo, Wicksell or Keynes' effigy appears on the currency of a country or monetary union, we believe that at least in terms of intentions, achieving sustainability through money is a concern. We also think this way if we see their picture next to (or instead of) the picture of governors of central banks. And we think the opposite when scholars perceive the natural rate as a nebula.

In a synthetic way, we learn, together with those whose names we mentioned, and not only, that the one who establishes the natural, normal dimension of the price of money is the entrepreneur, not the bank. The latter just follows the instructions; it receives the order from the entrepreneur, who, through the estimated rate of profit, communicates concrete data on the scale and duration of the "green cheese" production. The bank is invited to understand that if he subordinates its policy to the philosophy of the natural rate, an endless monetary increase does not make sense; it actually has limits. In addition, it is clear that if we reach the stage where economies are already absorbed by investment, labour is engaged and growth occurs without inflation, it would be good to quench its quantitative appetite. Or put an end to it, and, if possible, withdraw its "offspring" if the expected average rate of profit is declining. That is, the liquidity trap is the work of the short-sighted, of proof of logical indeterminacy. Whom should we give money to if production yields are declining? Okay, you can throb with injections to "stimulate" business appetite. But with what? You have to own the money you are claiming interest on, Turgot warns! If money is made not to measure but only to mediate, it is true, the idea of ownership is relativized. Is it a serious indictment to say that we can lend something that is not ours, thinking about Menger's "means of exchange"? Wicksell made it clear that it was one thing to operate with savings and sacrifice the present, as this will not bring about inflation but the prospect of developing, gradually and healthily, without speculative bubbles; and there was another thing to borrow by getting loans or running money out of the press, as this would get you to a dead end. You have big but "forced" savings, with little chance of being absorbed by the real economy. And then? Then ... you start purring about a possible secular fall, rejecting the truth, pretending to forget that the bank, the "hen", took out chicks without eggs, which it sends to the world, without anyone expressly asking for them and that, until they return home, or break like balloons, time goes by causing turbulence.

There is something else to remember here. One idea is that the expected rate of profit becomes natural and turns into a natural rate of interest only in conditions of free competition, without monopolies. Dominique de Sota's "injustice of monopolies" can ruin the whole business. In the case of monopoly, that is, we have neither natural prices nor the natural rate of borrowed money. Historically, however, it was possible to set the interest rate without competition. It was possible in an economy where goods are sent to the market with predetermined values and prices. In such a context (we are thinking about a planned socialist economy) the interest of the Central Bank is legitimately and logically entitled to the status of "guiding rate of interest". We see that the name given to the reference interest rate using the designation "guiding" betrays something. It says something about the emphatic title of the Central Bank and its claim to be first violin and court of justice. To its perplexity, within the perimeter of the free market, the philosophy of natural rate of interest grants it another status. Here, the bank is providing for the real economy, the entrepreneur; not the other way around. Even the bank's behavioural rationality is a function of the entrepreneur's behaviour. The idea of "independent authority" is at war with logic. If a bank issues banknotes with the effigy of Wicksell, it "unconsciously" admits that it is a part of the chain; that everything it does is subordinated to the real economy where the entrepreneur is the lead actor.

Then there is something else worth remembering, something that was said even before Wicksell. Namely, every participant in the economic activity is remunerated according to the "judgement of an honest man"; that is, they receive "nothing more, nothing less" than they deserve if they accept and respect the rules of the game. And the one who is responsible for things to go this way is not a "Controleur-génial", but the free market. The natural price is formed on the free market; it is a resultant in which all the natural prices of the participating factors flow into. The natural price of money is no exception to this logic. Smith and Ricardo's estimated profit, or the natural rate of interest in Wicksell's terms, is and it must be a resultant if it claims to naturalness; we are talking about a profit obtained in conditions of free competition at all stages of the production process and all the factors involved, i.e., labour, capital, land, etc. By no means a monopoly profit. Only thus conceived, as a resultant of a mechanism that operates with natural prices for all factors, can the rate of profit be called the natural rate. And when it receives this name, the natural interest rate justifies it as an expression of a "possible experience" in the Kantian sense. That is, an experience that bears in itself the signs of naturalness because it leads to a process controlled by the free market in order to produce natural, fair results. As an "a posteriori truth" the natural rate is set in front; it is a premise. It is established *before* the nominal (bank) rate; it goes through an adaptation path, by a Walrasian tatonnement process "slowly and with considerable hesitation" trying to find the light, never forgetting the forerunner. And the trailblazer is the entrepreneur.

If this is the game, what is the responsibility of the Central Bank? Assessing the productive purposes of those who borrow, as Böhm- Bawerk puts it, it must open the window and look at the market to find out, as Keynes advises, what is the business investment rate, its evolution in time and space. It does not need sophisticated econometric calculations to find out what it has to do. The "guiding rate of interest" comes from the outside, from the real economy. The game it plays is drawn by Ricardo and polished by Wicksell. From Ricardo it learns that the requests depend on the comparison between the expected rate of profit that can be obtained by using the borrowed money and the bank rate for which it is willing to lend. And the bank must be willing to lend because otherwise it no longer justifies its reason for being. From Wicksell it learns all the three hypostases in which, just like the entrepreneur, it can be part of a dynamic economy and in which the natural rate gives the signal: (a) Bank rate < Natural rate: inflationary growth, investments higher than voluntary savings, rising prices; (b) Bank rate > Natural rate: deflation; (c) Bank rate = Natural rate: phase of temporary equilibrium, savings absorbed by investments, growth without unemployment and inflation—the golden triangle of the classicists. It is not abnormal, we also learn from Wicksell, for the natural rate to be negative if the profit expectations of entrepreneurs are pessimistic. In other words, we have a theoretical framework in whose registers we can find out what to do in any kind of situation. It is important for the Central Bank to "keep an eye" on the economy and, depending on the signals received, to regulate its own policy. Conversely, if it takes control of the "lighthouse", the game it enters and which it controls can become a game with money for money; money out of nothing, producer of strange money, toxic and deceptive bubbles; denominated in "purely" monetary or "purely" banking inflation!

Apart from competition without monopolies, the natural interest rate is doing its job only on grounds of "honesty", fairness and the necessary possibility to anticipate results. The business of money lending separates, in time, the act of lending from that of repaying the amount due. A circumstance in relation to which the bank sets the duration and conditions of the repayment; a circumstance that forces it to think about the potential solvency of the debtor whose profit, in case of an economy where the central bank's budget is part of the general budget, is a source of its own gain; and, finally, a circumstance in which the borrower must be very good at anticipation. If borrowers let the bank inform itself in their place and make forecasts, we can no longer talk about the lever function of the natural rate; the bank becomes a pseudo-entrepreneur and the game moves exclusively to the court of the nominal economy.

Finally, another idea is worth remembering. Holding monopoly over money issuing, the Central Bank is "getting sick" and steps out. It becomes, just as Ricardo predicted, abusive, taking advantage of the fact that the government needs its money. Then it lends money to the government, subordinating it and it forgets that the just price also refers to the public interest. It makes its own forecasts, anticipates and turns the "reference interest rate" into a managed price. It becomes a court; a last resort to whose "solidity" the state is called upon to humbly make its contribution. From mere instrument, it becomes the "queen of the ball"! In Aristotelian terms, it becomes an "unmoved mover"; everyone is at its beck and call, tormented to decipher its cryptic messages. And, just as the Pharaoh set the level of taxes according to the flood rate, the central bank sets the interest rates according to its desired rate of sustainability. Thus, the idea of natural rate becomes an abstraction. An impossible scenario, we are entitled to say. And yet! We will find that it is possible to see a Central Bank reluctant to the "judgment of an honest man".

3.3 Post-Wicksell and Postcrisis. Current Metamorphoses of the Natural Rate of Interest

The natural rate disciplines; it sends you to the job description and shows you the limits of your competence. Who likes that? Definitely not the Central Bank and the Government. They enjoy their role of last resort. Or, the natural rate does not support such a thing. It listens to and follows a single rationality, the one that comes from the market. And then, what do we do with the natural rate when we claim that those who do not have a functioning market economy cannot claim to belong to the civilized world? We compromise. We let science do its job. We make it clear that sending "heretical" ideas to the normative is only possible with permission from the masters. And the "masters" do not need doctrinal heresies. They know what to do. In addition, they have their own flatterers who sign their payroll.

In such a landscape, what happened to the natural rate after its theoretical initiators disappeared? In short, the natural rate has not disappeared, but it has been shaped and transformed according to the needs of the moment. It has received new contours, facets and interpretations that push it away from its original meanings.

Such a route was facilitated, as we have tried to suggest, by the ideational environment in which it tried to find its sources; a troubled, contradictory, questioning environment unlikely to support the takeover and preservation of such a deep and healthy idea. We are talking about a context where anything is supported and argued in all directions.

It is not easy to grow, perpendicularly, on a field full of shifting sands. However, there are many attempts to support the idea of natural rate. Several initiatives claim to be ground-breaking. We mention here, Michael Woodford (2003) with *Interest and prices*, John Taylor (1993) with *Discretion versus policy rules in practice*, and Laubach and Williams (2003) with *Measuring the natural rate of interest*. Those who swarm around the ideas aroused by these works claim to be either Wicksellians, or Neo-Keynesians, or both. On a general note, they all place Wicksell at the forefront. Dismissing Smith, and especially Ricardo, is more than damaging. The background analysis, lucid and crystal-clear, belongs to Ricardo. Wicksell builds on it, gives a name to the phenomenon, but he is also unclear at times. Referring to Ricardo would spare us the interpretations; this is not Wicksell's case. That is why today's and yesterday's authors allow themselves to redefine the phenomenon, or put forward bizarre interpretations. And, doing so, also in a general note, they send the problem to areas where there is no trace of Ricardo or Wicksell! And there, they ask themselves absurd questions with even more absurd answers.

Concretely, Woodford raises curiosity through two works. The first work written together with J. J. Rotenberg (Rotemberg & Woodford, 1997) is the popularization of a methodology he relies on in his 2003 neo-Wicksellian paper (Woodford, 2003). Dreaming of a cash-free and with pure credit economy, framing the natural rate variable into a neo-Keynesian, dynamic stochastic general equilibrium type model, he deduces that this is an equilibrium rate, belonging to an ideal economy with completely flexible prices and which allows the equalization of potential GDP (obtained in the absence of any price rigidity) with aggregate demand. It is true that Wicksell was concerned with the status of the natural rate as equilibrium. But for him, as for Ricardo, this was a phase derived from the basic concept: the natural rate of interest is the reflection of the expected rate of profit. This is where we start. It is also true that he also mediated judgements with causal relations between the nominal rate and the price level. But, if prices are related to anything, that does not mean that by dealing with prices we automatically become neo-Wicksellians. It remains to be seen whether comparing the potential GDP with aggregate demand is congruent with the ideas of the one who put it forward. In another version, closer to the founders but still based on the equilibrium, Woodford turns the natural rate into a lever which makes it possible for the economy to enjoy full employment and stable inflation; provided that prices and wages are flexible. Judging by his assertions, Woodford is a Wicksellian; by condition, he is Ricardian.

The fact that Woodford's revolution does not break patterns through absolute novelties is also proven by those who interpret him, thinking of Wicksell. Here is what Adrian Penalver, a researcher at the Central Bank of France, says: "according to Woodford's (2002) seminars ... the natural rate of interest is the real rate that avoids rising or falling inflation if prices are fully flexible" (Penalver, 2017, p. 1). Do we have another version of Woodford here that is entirely foreign to Wicksellian philosophy?

Th. Laubach and J. C. Williams (2003) carried out estimates of what is believed to be unobservable but interesting, such as the natural interest rate, the output gap or the natural unemployment rate, in their capacity of "advised interpreters" of Woodford and Wicksell. Jean-Stephane Mesonnier (2005) analyses the reckless attempt of the latter. It is argued that, with a Kalman filter and in a semi-structural approach they try to outline the medium-term evolution of the natural rate. To better serve central banks, the working definition of the natural rate of interest is seen as "a real short-term rate compatible with a zero-output gap and stable inflation over the medium-term horizon" (Mesonnier, 2005, p. 48). What useful information do such estimation attempts bring to central banks? Some discouraging ones, such as: (1) "the short-term equilibrium rate of interest fluctuates widely from its long-term average"; (2) "the equilibrium rate of interest in the Eurozone seems to be set on a downward trend since the late 1980s"; (3) "the estimated natural rate is surrounded by a total uncertainty" (Mesonnier, 2005, p. 52). The third conclusion overrules the first. The excuse lies in the "unobservable" nature of the natural rate. We believe that the fault is elsewhere: in a dizzying definition of the natural rate followed by attempts to search for it in the wrong place. Then, we argue that one needs some courage to aim to determine the dynamics of the unobservable natural rate of interest through the unobservable gap in production; the courage to endure the criticism of reason for the sake of compromise.

Laubach and Williams' attempt is, this time, purely econometric. Observable variables are causally linked to unobservable variables but, according to the authors, they are determinable. Each of the six equations of the model (Laubach & Williams, 2003) has its quirkiness. Again, the output gap is related to the gap between the observed real interest rate and the natural one. Econometrics tells them that the gap between actual and potential GDP is induced by the gap between the real, short-term interest rate and the natural rate. There may be a reverse of the relationship as well. But, more importantly, they start from the idea that the big problem central banks are facing is to estimate, simultaneously, the natural rate of interest and the production gap! Where does this great confrontation come from?! Nevertheless, called upon to solve the dilemmas of central banks, the econometrics of the two authors conclude sadly: science fails at the natural rate. The variation margin of the unobservable fantasy, in its name the natural rate, is much too high to recommend it as a guiding beacon. Prudence is better!

If we want to capture the whole extent of the mess that dominates monetary policy, presumably in search of the natural rate of interest, it is enough to look at just a few works or positions on the subject. For example, Lawrence Summers (2014) suggests, via the IMF, the possibility of secular stagnation. Bernake (2005) was afraid of an over-saving at world level that kept long interest rates low. Economists take the ideas and submit them for validation. And the premises and conclusions are not consistent. Hamilton et al. (2016) are, in turn, intrigued by the hypothesis of secular stagnation. They test it using the natural rate principle: stable inflation and full employment. And they do not get anything clear. The only certain thing they can rely on is the uncertainty surrounding the "obscure" natural rate!

But, by far, the truly obscure attempt, sending the monetary policy into the area of undisguised precariousness, is the famous Taylor rule. We do not believe that the author intended what he got in the end. It is believed that John B. Taylor, professor and expert on monetary matters, wanted to serve science, and his merits are unquestionable. Here we are interested in its well-known and widely used formula for its contingency with the natural rate (Taylor, 1993).

The premises Taylor starts from send his judgements to the works of Wicksell, Fisher and Keynes alike. The hypothesis that inflation is inversely related to the difference between the nominal and the natural interest rate is "dear" to Wicksell and even Ricardo. The premise that the Central Bank can aim at inflation, manoeuvring the nominal rate in relation to the real rate, is in the ideational field of the mentioned economists but, especially, in the range of wishes of the Central Bank. The idea of the real interest rate, as a constant difference between the nominal rate and inflation, comes from Fisher. The principle according to which changing the real rate means changing the nominal rate more than inflation can legitimately be associated with Taylor's name. Step by step, the author reaches the following formula:

$$L^* = r^* + p + a(p\ p^*) + b(y\ y^*).$$

The formula is meant to serve the bank and, therefore, we need to identify the factors the nominal interest rate L^* depends on. As it can be seen, this mainly depends on the natural rate, or the equilibrium rate r^*, the observed increase in prices, p, the gap between the inflation targeted on the short-term by the central bank and the actual inflation pp^*, the gap between potential output and the actual output yy^*.

Trying to adapt Wicksell's theory, the formula basically suggests a positive relationship between the economic activity (production) and prices (inflation). Hence, the implicit conclusion, and the ideal desire, that when you aim at, and succeed in stabilizing inflation at p^* level, you succeed, *ipso facto* in stabilizing the activity at y^* level. How easy seems the task of the central bank. But is there such a "divine coincidence", Cristian Bordes (2008) wonders? We ask ourselves the same thing; and we doubt it; and we are not the only ones.

Beyond our doubts or those of others, the formula encompasses causal relationships with unhidden, fundamental flaws that undermine its credibility. Broadly speaking, it sends us to the field of a battle between the living and the dead, between *ex ante* and *ex post* parameters, from which the only thing left is an invitation to chaos. Thus, for example, the natural rate is linked to the relationship between the level of full employment and a target inflation rate. Do we wait for it to happen and observe something like this, and then calculate a nominal interest rate for a period to come? In this context, Taylor does not tell us that this rate has anything to do with the expected rate of profit. Then, setting the nominal interest rate based on the difference between what you aim for and what you notice in the dynamics of inflation is extremely bizarre. Targeted inflation expectations are those of the central bank and they are related to the free market just as the natural rate of interest is related to the profit estimates of entrepreneurs.

In addition, all imperfections known to determine the inflation index spill over into the formula. If there is something more to add here, it is the fact that the Central Bank does not know in advance what is the size of the monetary expansion produced by those on the large side of the inverted pyramid, that is, by the commercial banks. It finds out, ex post! The production gap is no less problematic. Even though Fisher is present here, his deflator does not give more credibility to the formula. The difference between a potential GDP and an actual one is not one between physical parameters. It is true, intermediate consumption, import-export balance, etc., can affect, physically speaking, the calculations. But the important thing is that all the imperfections induced by the dynamics and structure of prices come up here. We have prices for capital goods and consumer goods; goods that are the target of the free market or state-managed prices. Is the Central Bank willing to assume a formula that seeks the pivot of its action in the wagon of uncertainties and imperfections known to accompany the determination of Taylor's imperfect GDP? In addition, the complicated arithmetic of potential GDP comes up. Even with a clear definition, even deduced from the equilibrium area in which the main factors of production, labour and capital are used to their full level and remunerated at natural rates (here Ricardo and Smith should be consistently mentioned), even so, the studies carried out on the subject show that it is not an easy task. But let us not forget, here as well, the actual GDP is an ex-post parameter; in relation to a nominal natural interest rate that claims to be ex-ante!

Overall, this formula used by central banks shows, relevantly and undeniably, how a key institution of the free market can break away from the free market. It seems to be the work of a monetary planner animated by the intention to subordinate the economic dynamics to a conductor who likes anything but being an instrument. Because, what does a Central Bank do with such a formalized rule on its table? For the health of the economy, it would be best to circumvent it. But it does not. The academic anchor received from the highest intellectual circles acts as a spring that sends it to the area of eminence, tickling its pride and making it take its role of last resort way too seriously. What entitles such remarks? The fact that the factual exercise, vividly captured in specialized literature on the subject shows that this is the case. Central banks have long forgotten their primary role, to ensure the security of the payment system and feed the credit pyramid according to clear rules, consonant with the spirit of the free market and contrary to usurious practices. Their job descriptions include everything: inflation targeting, price stability, financial stability, economic growth surveillance, etc. The "Minsky moment", the debt crisis and financial fragility pushed it to replace the finance ministries. Anyway, its permanent position is that of supervisor, of controller. Up to the point where the following question is perfectly legitimate: what does the government do then?

The fact that the same monetary policy is successful in one area and not in another can also say something about the reaction of the business environment to the monetary policy. A good example is the success of the Friedmanian monetarist policy in 76-88s Germany and its failure in the USA. The American stock market-like capitalism was compatible with Robinson's recklessness of bank expectations. In Germany, the "culture of stability", the direct involvement of banks in business, the strong institutionalist aura of money and the secular fear of inflation do not allow playing with money. There the government does not let the bank do what it wants.

Faced with nebulous formulas, it is not surprising that the Central Bank makes decisions in conditions of uncertainty. Despite what it believes, uncertainties do not come from competition and the free market. They come from its own philosophy of monetary policy; confident in its own calculations and expectations and distrustful of market information. If it were not so, Taylor's formula would not look like a long list of determinants in the estimation of which no one except for the Bank dares to take part. And where the output gap—that it calculates as well—seems more important than inflation. The first and most important factor, the natural rate of interest, suffers a sharp decrease in weight, compared to the other components of the formula. And, it is worth remembering, the size of the natural rate is not the result of a dialogue with the entrepreneur. In bank arithmetics, this is a secondary character. Or, if you dilute it, if the expected rate of profit is not the one that sets the tone, the natural rate no longer functions as a wand. The hope for sustainability is in vain. Insisting on the natural rate, Ricardo, Wicksell and even Keynes provided theoretical registers for all situations, normal and less normal. The hypothesis that the scholars mentioned did not prepare us for risk management does not hold. In Wicksell's logic, risk changes the natural rate; high risk means anticipating problematic profit. If a small natural rate "sticks" above the nominal rate, it is not the fault of reality. It is the bank's fault that it does not know how to conceive its lending terms—in such situations—so as to avoid the insufficiency of the actual demand accompanied by unemployment. Keynes would know what to do. But it only thinks about its own profit. Forced savings, in addition to the incentive to risky investments, do not affect it directly either. Nor is it disturbed because its relationship with the natural rate, i.e., the entrepreneur and his profit interest, is separated by a curtain. When it comes out or is called to report by some parliament, its sincerity, craved by Hayek, does not breathe out. It says everything and nothing at the same time, by means of a sibylline stylistics. In louder terms, prudence—acting as a prima-donna—says what it wants or what those that fuel its wellbeing want to hear. The Bank of Norway (and the Scandinavian banks in general) is given as a positive example of its good faith towards customers; it announces the target rate, the production gap and, if possible, the natural rate. By contrast, the Bank of England does not publish anything. Because it "doesn't know"! (Bordes, 2008). But the Bank of England knows something! It knows it cannot announce what it doesn't know; it will discover the production gap and the inflation gap together with those interested!

3.4 Instead of Conclusions: Why Does the Central Bank Refuse the Sustainability of the Natural Rate of Interest?

Critical analyses of the central bank show sufficient circumstances in relation to which it appears as an indispensable yet unpopular actor on the economic stage. There are few (if any) instances on the subject that accuse it of the fragile relationship with the institution of the natural rate. And this is exactly how the relationship is: crook, perverted or malformed. And why does the Central Bank run away from the natural rate it places in the world of unobservable fantasies or abstractions? It probably has—we would say—its own reasons.

First of all, the Central Bank enjoys the status of master, controller, supervisor and conductor. And this directly and not indirectly. It enjoys interfering directly with the money supply, prices, inflation, economic growth, etc. Or, the natural rate is an indirect lever. A lever which would tie it, umbilically and to the same cord, to the entrepreneur. And, on this occasion, it would tie it to the real economy. This would affect the desired status of distinct identity, specialized in things that are inaccessible to the common people. The natural rate would simplify the job description by compressing it. Its only job, as Ricardo and Wicksell say, would be to look at the market; contact the entrepreneur and receive information about the expected rate of return. And, depending on what the entrepreneur requires, to provide the necessary liquidity at a loan rate that does not cancel the entrepreneur's impulse to invest. That is it. However, doing so would basically mean acknowledging that the wand is in the hands of the entrepreneur. This is where that "first" comes from, followed by "then", the loan rate seeks its way, slowly, stumping, but still seeking. However, it wants this "first" in its yard. It is the one that sets, "first", the "guiding", reference rate. And everything follows, stumping, from here, on an opposite path than the one conceived by the mentioned founder. The very meaning of the information flow is established from the bank to the market, to the real economy; its information versus market information. Its information and estimates are better; the evil and rebellious markets can only provide distorted information! Its "friendships" cannot be there. The natural rate would make it a sibling of the entrepreneur. Or it wants to be a sibling of the government. Setting foot in its yard, the bank takes over the prerogatives of the finance ministry and handles the financial crisis. Anointed by the government, it can manage anything: inflation, debt, exchange rates, cartelization or the crisis. And, above all, bankruptcy only occurs with its permission. Running away from the economy and cloistering in its own paradigm, the bank does not forget its aura. It cultivates it with a warm, thoughtful speech abounding in formulae. It does not hurt to recall that there is a natural rate. Through it, the bank keeps in touch with the scholars. They provide formulae in which the natural rate is present; epsilon weighted! The circulation speed of the money supply or the outstandingly important reserve fund do not belong in the formula. A formula that, calculated—we do not wonder how and why—turns the Bank into a reliable guide. Until the next crisis.

Secondly, in its attempt to refuse the ungrateful role relegated by the natural interest rate, the central bank gets help. In vain would it look at the market because it would not see the right thing. It should find the average rate of profit there. Or the average rate of profit is not calculated; neither on total economy, nor on its branches. The bank does not ask, the statistics do not provide! The poor bank has to do its own calculations; to anticipate, inform and call econometrics to help it impress. Wrap it in formulae and make it impregnable and, forever, indispensable!

Third, contemporary monetary theory supports the bank's attempts to escape from the real economy and only bathe in its own waters. As long as, everywhere, it is stated that there is no consensus on the natural rate, potential GDP or inflation, the Bank may, in turn, declare that it has nothing to support its policy. The theoretical registers that could support it are slippery. And, unfortunately, its opinions are confirmed. As we have already noted, developments on the idea of the natural rate are either missing or pale precisely in the works of those who set out to demystify the role of the central bank. The harshest reaction is from the Austrian School. But diluting the role of the natural rate in the mechanism they want to dismantle reduces their discourse effectiveness. We know that Hayek mentions the natural rate, and correctly calls it equilibrium rate but, in no more than two sentences, he dismisses the problem. How effective his speech would have been if he had taken into account a complex of natural rates, based on which entrepreneurs could make decisions regarding whether or not it was worth moving to the next stage of production. His fellow scholars only touch upon the natural rate among others, without further discussing the subject. Too bad! The ethics of money production would have found another support. Hülsmann wants to get rid of the "Brussels Moloch" but how? Dreaming of gold and not caring how it is possible and why does the central bank do what it wants. Rothbard does the same. The fiercest fighter against the Central Bank, the one who reveals its treacherous underpinnings, is not concerned with the main source of these institutional perversions: the divorce from the natural rate. Too bad again! As unparalleled defenders of the free market, they did not realize that getting the natural rate out of the game means cauterizing the most important figure of the market—the entrepreneur. In search of optimization solutions in the monetary area, admitting the circulation of several currencies, "firing the cannon" at the monopoly of monetary issuance, they were not concerned with the vocation of synthesis of the natural rate philosophy. And on this line, the necessary attention to the "invisible" potential GDP, capturing it in the touches of an algebraic sum of the virtues of the free market; seen as the result of the natural prices flowing into it, a process mediated by the competitive market suffocated by monopolies is missing. None of this!

What else is there to add? Radically different opinions about the control or impossibility of controlling the money supply? The existence or absence of a general price level? The need for or refusal of regulation? Inflation as a banking or monetary phenomenon? We know full well that interrogations can continue. But without changing the substance of the problem. And the problem is that the Central Bank can claim that it lacks the theoretical support to light its way. And then it lights it up on its own!

In conclusion, we have seen that the Central Bank has every reason to run away from the natural rate like the devil from the holy water. At the same time, we have seen that a change in its conduct is absolutely necessary. Is it also possible though? An echo from Böhm-Bawerk tells us that the rate of interest receives a natural level only through the involvement of successive generations. We agree. We also know that just letting the entrepreneur do the calculations, he will be tempted to take the roundabout way. If we leave it at the beck and call of the bank, there is a high chance it will turn him from an industrialist entrepreneur into a commercial or speculative one. In order to avoid this, the Bank must change its behaviour. We doubt that it will do it on its own initiative. But we have no doubt that the chances of sustainability will be severely affected in a key area of the economy if such a behavioural change does not occur.

References

Bernanke, B. S. (2005, March 10). *The global saving glut and the U.S. current account deficit*. Speech delivered at the Sandridge Lecture, Virginia Association of Economics, Richmond. Retrieved from http://www.federalreserve.gov/boarddocs/speeches/2005/200503102/default.htm

Bordes, C. (2008). IV. Au coeur des décisions de politique monétaire, la politique des taux et la coordination des anticipations. In La Découverte (Ed.), *La politique monétaire* (Repères).

Hamilton, J. D., Harris, E. S., Hatzius, J., & West, K. D. (2016). The equilibrium real funds rate: Past, present, and future. *IMF Economic Review, 64*, 660–707.

Hayek, F. A. (2012). *Law, Legislation and Liberty. A new statement of the liberal principles of justice and political economy*. Routledge.

Hülsmann, J. G. (2008). *Ethics of money production*. Ludwig von Mises Institute.

Laubach, T., & Williams, J. C. (2003). Measuring the natural rate of interest. *Review of Economics and Statistics, 85*(4), 1063–1070.

Mason, W. E., & Butos, W. N. (1996). *Classical versus neoclassical monetary theories: The roots, ruts, and resilience of monetarism—And Keynesianism*. Springer Science & Business Media.

Mesonnier, J. S. (2005). L'orientation de la politique monétaire à l'aune du taux d'intérêt «naturel»: une application à la zone euro. *Bulletin de la Banque de France, 136*, 41–57.

Penalver, A. (2017). *Taux d'intérêt naturel: estimations pour la zone euro*. Retrieved April 8, 2020, from Bloc-notes Eco website: https://blocnotesdeleco.banque-france.fr/billet-de-blog/taux-dinteret-naturel-estimations-pour-la-zone-euro

Rotemberg, J. J., & Woodford, M. (1997). An optimization-based econometric framework for the evaluation of monetary policy. *NBER Macroeconomics Annual, 12*, 297–346.
Rothbard, M. N. (1994). *The case against the fed*. Ludwig von Mises Institute.
Stiglitz, J. E. (2010). *Freefall: America, free markets, and the sinking of the world economy*. W. W. Norton.
Summers, L. H. (2014). Reflections on the 'new secular stagnation hypothesis.' In C. Teulings & R. Baldwin (Eds.), *Secular stagnation: Facts, causes, and cures* (pp. 27–40). CEPR Press.
Taylor, J. B. (1993). Discretion versus policy rules in practice. In *Carnegie-Rochester conference series on public policy* (Vol. 39, pp. 195–214). North-Holland.
Woodford, M. (2003). *Interest and prices: Foundations of a theory of monetary policy*. Princeton University Press.

CHAPTER 4

Social Pressure—The Risk of Invalidating the Lesson on What Makes an Economy Sustainable

The relation between economics and the social aspect of society has been and remains heated to this day. Large amounts of passion and ideology have been spent in this regard. As of recently, applying an environmental hallmark to this relation has made it volcanic. Several theoretical developments produced in this respect consolidate the status of economic science. At the same time, many, all too many professedly scientific creations, suffocated by propagandistic populism, produce opposite effects. In short, in the field of social matters, a bit of foolishness is all that is required to claim that one is creating economic science!

4.1 Can Social Tensions Shift Causalities and Undermine the Logic of Sustainability?

Regrettably, YES is the short answer to the interrogation in the title. We reach this unwished-for answer by following at least two tracks: (a) the egalitarian drift from the episteme of wealth production and distribution as conceived by the classicists; (b) the distorted interpretation of the so-called pre-eminence of the man-things relationship with respect to the man-man relationship within the plan of the founders.

Firstly, it is worth observing and remembering that the episteme of wealth defined the preoccupations of the founders as well as the object

I. Pohoaţă et al., *The Sustainable Development Theory: A Critical Approach, Volume 2*, Palgrave Studies in Sustainability, Environment and Macroeconomics, https://doi.org/10.1007/978-3-030-61322-8_4

of the science whose foundations they laid. A culture as a "locus of simultaneity" of the aim, the method and the instruments made to serve the production and the distribution of the result, as Foucault put it in a proper and inspired manner (Foucault, 2005, p. 182). The founders placed labour at the centre of the craft of achieving wealth and, subsequently, development. Its glorification rather than that of ease supported and provided arguments for their project. The idea that development must find its finality in both individual and collective good was also very clear and well represented. Their lessons, excepting Mill's wanderings, support the logic according to which collective wellbeing is reached by aggregating individual wellbeing, not the other way around. We are talking about a mechanism rather than a relation of subordination. And they understood that being part of the mechanism as well as partaking of general wellbeing is an individual affair rather than a linear experience. Everyone contributes as best as one's own different intellectual and physical potencies allow. The specificity of this participative act induces the idea of the specificity of everyone's involvement in the act of distribution. If wealth is achieved in a differentiated manner, the reward is different as well. From this point of view, the founding message was clear: any attempt to a-priori engineer individual wellbeing by transforming it into a drop in the ocean of society's wellbeing, engineered in the same manner, can only lead to undesirable results.

Until practice confirmed their apprehensions, there were more than a handful of messengers who confiscated their message. The first attempts occurred in a strongly institutionalized context. We are talking about the *Declaration of Independence* and the *Declaration of the Right of Man and the Citizen*. Both prophesize the privilege of natural equality granted at birth. If Americans thought about equality as a reply to unjustified privileges, calling for the "pursuit of happiness" (not promising it to everyone in equal amounts), France, the country where socialism originated, only admitted one exception to the innate equality: "Social distinctions may be based only on considerations of the common good!", imperatively stipulates its *Declaration*. Namely, the primacy of common good received state empowerment. It is no wonder that what followed occurred, both factually and conceptually, under the sign of Nature that generates equality of rights and goods. An extreme and vocal supporter of this idea, Babeuf thought that "[i]n a true society, there must be neither rich nor poor" (Birchall, 1997, p. 126). Babeuf, the leader of the "Conspiracy of equals", did not end well, but his idea did not die with him; conversely, it bred and fed such theories and doctrines whose end of the road is visible under our very own eyes. Thomas Piketty is not made of the same stuff as the above-mentioned revolutionary; however, he has plenty of ideas that would have complemented the known tribune's platform of destruction of any inequality and reestablishment of common happiness.

Nonetheless, it has to be remembered that, progressing through time from the Paris Commune, through the socialist utopians, Marx and all the "scientific" creators of communist works and practices, all these people, together with Lenin, Mao, Stalin and the others have sent towards us the phantasm of equality in the corrosive and utopian formula of absolute equality. This formula is the origin of all the public social pressure manifested whenever inequality becomes upsetting. In other words, we mean to say that nowadays it is not the Marxist perspective on exploitation and the unjustified appropriation of surplus value that drives people out in the street to protest. No, the foundation of social claims is not determined by an absolute evil, but rather by the revaluing, either at one's own initiative or "assisted", of the utopian form of egalitarianism. The improved capacity of control over the political process exerted through voting by the population alleviates some of the social pressure.

The Keynes moment, when an economist with liberal convictions, a product of Marshall's school, forcedly became so socially conscious that his very doctrinal position was seriously and forever called into question, supports our assertion. Back then, in 1929, the European working class, Marx's "industrial reserve army", would rather have been "exploited". However, the technical–economic conditions of production could not allow for new hirings, quite the contrary. And it was also back then that the lucid and rational Keynes realized that without social peace, with the streets teeming with unemployed people, there was no economic peace. He was already trained by the morbid and socially strained landscape shaped in the aftermath of World War I when, temporarily abandoning his original doctrinal position, he was writing that "[m]en will not always die quietly. For starvation, which brings to some lethargy and a helpless despair, drives other temperaments to the nervous instability of hysteria and to a mad despair. And these in their distress may overturn the remnants of organisation, and submerge civilisation itself in their attempts to satisfy desperately the overwhelming needs of the individual. This is the danger against which all our resources and courage and idealism must now co-operate" (Keynes, 2019, p. 176). Had he not signed these words, it would have been difficult for us to identify their author. Everybody knows today who wrote the famous sentences that let us know that it would be good, based on the verified principles of private initiative and laissez-faire, for the Treasury to bury and then dig up its banknotes as a "better-than-nothing-at-all" alternative to the necessity of unemployment rate reduction! We do not comment on the uselessness and the poor logic of the gesture. We are interested in this episode as it is one of the best illustrations of the way in which, under social pressure, economic logic no longer amounts to much. Thanks to this example, we learn that when the course of economic dynamics is diverted by harsh violations of its inner laws it is possible to open wide the gates of economic judgements supported in odd causalities. Keynes succeeds in making plausible an unproductive investment just to escape an imminent and dangerous evil: mass unemployment. And he does so by inviting the state to briefly "assist" private initiative. He placed great emphasis on social concerns while neglecting the lesson of the founding teachers. He called on the state to help and, without considering the consequences, transformed it forever into a simultaneously first-importance and last-resort employer. In essence, he converted it into a great generator of false jobs, hid under the protective garment of a just-as-false and precarious public happiness.

By placing the state at the beginning of the causal chain of economic and societal processes, Keynes left the de facto "good old doctrine". It is well known that the good old doctrine favoured the accomplishment of social interest by way of individual interest; it fostered private initiative in creating wealth and implicitly, of meaningful jobs; it put production at the beginning of the economic cycle as a logical phase from which everything derives; it made the state responsible for the security of the country and the citizens, a small but powerful state, initiator of an institutional arrangement that stimulates business appetite. On the other hand, what gate did Keynes' state interventionism open? One where the state is invited to be the depository of collective wisdom, where common interest and its promotion are shaped by electoral purposes; where distributive justice conquers the logic of production efficiency; where jobs created before starting production become first-hand electoral trump cards; where optimality and equilibrium cast out the individual from the economic game. In other words, we are talking about a large, omnipresent and omnipotent state, in charge of everything: hospitals and schools, roads, air temperature and tax rates also. A state that lends itself to criticism when raising taxes, but is also worth embracing when it creates jobs. It is capable of serenely and "logically" explaining everything that happens within its territory, which is vast and full of upside down truths. This is what happens when the state is allowed to play the part of the master and is not subject to the merciless institutional yoke called for by James Buchanan.

Keynes' transgression against classical precepts on the efficiency and role of the state and shift of the economic dynamics and sustainability analysis' focus on the criterion of full occupation gave a strong signal. It is the beginning of the dominion of social considerations over economic matters in assessing sustainability. From Keynes onwards, economists, and not only them, would use the ideas of the most quoted representative of their profession whenever they felt the need for a famous support in overthrowing well-known classical causalities.

The environmental footprint enjoys the same status as a strong source of pressure on standard judgements. It is inappropriate to balance severe social problems in moments of large economic crises with those related to the heavy deterioration of environmental conditions. There is no way for the issues of the environmental footprint to pervert the essential causalities of economic dynamics. Their specificity does not allow for such openings. At the same time, a "distortion" does occur. By invoking the specificity of the domain, the scholars concerned with it deny the laws of standard economic science in explaining the phenomenon and surmounting its asperities. They want another or other science(s). They do not want to sever the umbilical cord to the mother science; however, "something else" would be more appropriate. And so, we can see how, under the pressure of social considerations, economic science is capable of delivering phrases that verge on the ridiculous, while under the pressure of the environment, it risks breaking into pieces.

Secondly, it is not only the inclusion of social or ecological considerations to ensure balance and harmony that may encroach, by inadequacy or wrong formulation, upon the logic of an economy lesson. The general accepted opinion that economists know a lot of economic science, but nothing else, also contributes to this situation! Ethical, moral and psycho-sociological matters, as well as the general understanding of human nature, may elude them! Economists are good at efficiency calculations, rentability, factor substitution and determination of marginal efficiency yields, the preparation of budgets and the approximation of price evolution and exchange rates. A sort of day-to-day book-keepers, and not even perfect at that! If they were perfect, their econometry would not fail in predicting crises.

Of several allegations that may be brought against economics and the economists, the one mentioned above is the most unfair (unjustified). The history of this science and especially the status and the stature of those serving it are evidence in this respect. From the start, economics has been open to social concerns, and the scholars who launched it were very socially conscious. Actually, the very title of economist appears much later than the beginnings of this science. In addition, and in its defence, the good old tradition of entering the gates of economics coming from other disciplines such as history, philosophy, psychology, mathematics and law is not lost. Methodological, epistemological and ethical debates go hand in hand with the academic production of science. The *Theory of moral sentiments* finds worthy replies in works such as *Human action* or *Law, Legislation, and Liberty*, to give only two brilliant examples which prove that economics has been and remains a science with strong emphasis on socio-human aspects. One that has not forgotten man, understood in the plenitude of his more or less rational defining attributes, with his aspirations to a better life, both materially and spiritually. Not even the "arid" neoclassicists forgot this. Equilibrium and optimum were not developed while ignoring social concerns, quite the contrary. After finishing their abstract scientific construct, they came down to earth and looked into distribution and its equitableness, the expediency and advantage of cooperatives and the situation of women, and the possibility of combining the efficiency of capitalist production with an increase of fairness brought about by Walras' socialist coloured distribution. What else could they have done? It is true, their construct operates with several abstractions and with identical individuals, with homo economicus cast as its main character. One which cannot be identified as a specific individual, but it is still present in the form of an idea. It is the idea of a behavioural model of an individual who must be efficient in a world with scarce resources; a model of analysis specific for the economist just like the atom is the specific model for the physicist and the molecule for the chemist. As an entrepreneur—this is the de facto name of this character—he is interested, first and foremost, to earn profit. Without profit, his initiative is dead, and all those connected to it fail. Does he get rich this way? It is only logical! Yet, when he is too successful, social pressure makes its presence felt again. In its name and in the name of a kind of peace for which the spectre of egalitarianism defines the path to follow and, like a spider web that catches any transgression, it dominates and condemns any too obvious success. The economism, productivism and egoism of the character that creates inequalities because he gets out of line, as well as the scholar that advises him, the economist, must be brought under control. In the name of social peace!

4.2 Economism and Anti-Economism—Ideologies at the Borders of Sustainability

4.2.1 *A Brief History*

Economism has a complicated history. From the start, it was misappropriated by both left-wing and right-wing ideology. As a motive of political and doctrinal dispute, economism did not break with its original roots. Which one plays the part of the first violin: economics or politics? Which is the one to provide a starting point for a socio-economic construct? These were the questions both in Lenin's time and afterwards. Concerned with the relationship between economics and politics, Lenin (1951, 1961) weighed the radicalism of politics against the "cozy" character of economics. "What is to be done?" he wondered. Where does one begin? When he wrote *Left-wing Communism: An Infantile Disorder* (Lenin, 1900), he was pondering on the Trotskyist impatience of Mao's Chinese for building communism. He realized it was not the revolutionary and political radicalism that yielded the expected (awaited) results; the emphasis on the revolutionary modernization of the economy was appealing to him. What he meant by industrial and agrarian revolution and what he achieved are well-known facts. We are concerned here with the subject of strategy, with the matters related to the steps that need to be taken; with that which he thought should be firstly valued. And he first thought about the economy. There was starvation in his country. His "economism" did not yield results, as it was encapsulated in the non-productive logic of the red phantasm. Yet, the idea that one must first eat and have shoes to wear and only then make politics was here to stay. In its strange way, Leninism was also a first form of manifestation of economism, not as an exaggerated emphasis on economics, but as recognition of its pre-eminence. Looking at things from a historical perspective, we are compelled to observe that in the competing country, China, it was not Mao who followed Lenin's thought, but his successor, Deng Xiaoping, who proved to be an unequalled master strategist. He did not let politics stand in his way. He did not put socialism before the economy, but the other way around: this was his creed. It was a creed expressed in a novel formula: one country, two systems. In fact, China is a country that opened itself to the free market and allowed the communist structures to get tired and die alone of physical and moral erosion, helplessly witnessing how goods fill the economy and communists become billionaires. It is a defiance not only of the leftists, no quotation marks added, but also of everyone who saw economism as a deceitful and hard-to-defeat enemy.

What would Alber Jay Nock have to say today with respect to the fantastic display of China as the world's workshop? He thought that "Economism can build a society which is rich, prosperous, powerful, even one which has a reasonably wide diffusion of material well-being. It can not build one which is lovely, one which has savour and depth, and which exercises the irresistible power of attraction that loveliness wields. Perhaps by the time economism has run its course the society it has built may be tired of itself, bored by its own hideousness, and may despairingly consent to annihilation, aware that it is too ugly to be let live any longer" (Nock, 1943, p. 147). This is beautifully said, but also wrong—it has double resonance and double address. One is meant for the West which is "getting bored" of its own successes; however, it is neither hideous nor discouraged by everybody's share of material wellbeing. Economism would serve as the beating stock for Latouche, Piketty, Krugman & company, convinced as they were that wealth is achieved by placing emphasis on the economy, yet this alone does not lead to a wonderful society. The other resonance portrays economism as an emphasis on economy and the deliverance from ideological phantasms, which helped the Chinese overcome the starvation of the "Great Leap" and, gradually, eat three meals per day, fill the world with goods and export food. Should we ask them whether this is delightful or not? Or are they beginning to despair when, free to move around the world and buy whatever they desire, they have forgotten that to be happy during Mao's time amounted to owning a beret, an overall and a pair of tennis shoes?

We could continue with such questions without changing the answer. And the answer is that economism may be seen from various perspectives; it gains meaning depending on the specificity of place and time. But, basically, irrespective of the ideology it is injected with, it amounts to the same thing: the crucial importance and the domination of economics, its primacy in the equation of life; the idea that everything starts with economics and without it one's prophecies are for nothing. This idea defies and causes reactions when unwanted collateral effects become meaningful in defining individual and collective human wellbeing. It stirs reactions when profit, productivism, hedonism and material wealth pervert minds and alienate the human being from its normal life, when he sees nothing else besides economics and profits. Anti-economism is the name used to describe the reaction to such a situation. This reaction is normal and adequate up to the point when the critique becomes criticism. From there onwards, it becomes critique for the sake of critique, denying and casting doubt on the fundamental causal economic relations.

This brief history of the idea may help one better understand, today, why sustainable development finds contingencies in the concerns related to economism.

4.2.2 *New Characteristics of Anti-Economism*

On the path we set from the beginning, that of deciphering the message of the Brundtland Report, we agree that sustainable development, as theory but especially as practice, faces at least four challenges: (a) a demographic constraint—in knowing how much population can planet Earth feed under normal conditions; (b) a technical challenge—in finding out the maximum level of production that can be reached, under given conditions of technical progress, without disturbing the ecological equilibrium; (c) a Paretian confrontation—in deciphering the extent to which better results may bring satisfaction for one generation, but also within the relation between generations, in regard to both the topic of incurred sacrifices and that of equity or compensation; (d) a confrontation with the ever-present equation of economic efficiency—in finding the line that reconciles the need of effectiveness in profitably propelling economic dynamics and the socio-human finality of sustainable development. It is noteworthy that the four big problems to solve share a common axis. The solution to all of them relates to production, not one achieved every which way, but efficiently, in order to satisfactorily feed the entire population while shielding the environment and fully satisfying the human being of today and tomorrow. Things seem revoltingly simple. Nevertheless, the idea that starting from production, from economic growth, can allow us to also solve the other issues is disrupted by acclaimed points of view that induce the idea that things are not quite what they seem. The theory of economic degrowth is the most promoted and invasive example. On such grounds, we will reserve a distinct treatment for it. Going forward, we will deal with the other aspects of contemporary economism and anti-economism.

We take advantage of a clarifying landmark about economism provided by a non-economist, James Kwak (2017), in his book *Economism Bad Economics and Rise of Inequality*. The author says that economism is and, at the same time, is not an ideology. It is inasmuch as, using Leibnizian logic, it tries to persuade that today's world is *The best of all possible worlds*, any modification leading to worsening life conditions for all members of society. It is, in other words, an apology of a status quo that considers inequality natural. This is the critical target of the book. Kwak says that economism is not an ideology in the classical sense of the term because it does not rely on a separate doctrine, it has no political programme and no fervent devotees. And, if it is not an ideology based on its own epistemological foundation, then it must necessarily be a method for the interpretation of the economic and social world. And how does it interpret it? It does it in a blunt, reductionist manner, by means of an "interpretative lens" on whose basis facts, just some of them, are idealized and captured in abstract, astounding models and schemes that are, nonetheless, out of touch with the real world. This is considered "bad economics"—the abstract science of the neoclassicists and their neo-neoclassical supporters, a science of simplifications and "heroical" hypotheses, of idealized and frozen images about economy and society. A science that appropriates the university curricula, inoculating in the minds of students the image of an orderly, geometrical, articulated and untouchable economic system! The daring Nobel-prized beginnings of searching for the empirical evidence of abstract hypotheses are welcomed by Kwak. He mentions George Akerlof, Michael Spence and Joseph Stiglitz for showing how information disparities hinder the results of the exercise of the free market. Robert Schiller is remembered for proving the effect of investor's irrational behaviour on financial markets and Elinor Ostrom for revealing the role of institutions in solving problems that the market cannot solve, etc. In other words, it is suggested that such examples have to be imitated. How to do so? By means of a "better economic analysis"—by abandoning the simplifying schemes of neoclassical economism and tying economics to economic history, institutionalism, law, behaviourism, etc. We would need, in Kwak's terms, a new scientific story!

As far as we know, "the great story" has long been over and without substantial results. Reactions against the cosmopolitism and the abstractionism of the classical and neoclassical schools were given names that have become part of history: the German historical school, American institutionalism (the old institutionalism), economic nationalism, the reforming interventionism à la Sismondi and, not to forget it, socialism. The most scholarly and well represented is the German historical school, in both of its versions. It suffices to read *Political economy and its methods* and we find out that Schmoller was complaining about the same issues as Kwak. The same can be said about Hildebrandt or Wagner, who desired a "science of the laws of the historical development of nations" and respectively a "science of economic phenomena". For all of them, the path to economic knowledge had to pass through history, law, sociology, ethics and institutionalism. In his work, *The foundations of political economy*, Wagner complements this discourse and offers a solution as well: a socialist state in charge of social justice and the reduction of inequalities by means of income redistribution. A payer of indirect salaries is what James Kwak would like to have as well. This is the problem that troubles him more than it did the German historians and all the doctrinaire adversaries of classical and neoclassical liberalism—inequalities. The anti-economism manifested through the fury directed against models and modelling is just a mask, a derivative product; it is anger channelled against an instrument that cannot explain and clarify disparities of wealth and social status.

Leaving aside the fact that the critical arguments and the promise of a new success story lack novelty, Kwak provides us with two essential points for "our story": the doctrinal eclecticism of the scholars that bear a grudge against economism; the essence of the movement consists of the attempt to disjoint and relativize economics and everything connected to it (accumulation, production, profit, GDP, etc.) in relation to other sides and aspects of social-economic life. Thus, the line of conduct suggested by Kwak was and has remained the same: the workhorse of anti-economism has been the neoclassical model of the free market with profit serving as the main catalyser of all energies.

How much justification is there to be found in the ever more influential and virulent discourse of the anti-economism? And to what extent do its deviations from economic logic and from the declared goals of economic science have anything to do with the fundamentals of sustainability? Without trying to establish a mathematical relationship between the pros and cons resulting from the ever more obvious intrusiveness of economism, we affirm that the answer to the above question can be found in the interval between not a whole lot and very little.

We are convinced that the scholars who support this movement are genuine and well-intended. By laying their arguments on shaky foundations and by drawing conclusions against the natural logic that considers production rather than distribution as the beginning of a sequence of moments that converge to consumption, they sabotage their own goals. This is the story we are now dealing with.

4.2.3 GDP Statistics—Between Economism and Anti-Economism

Nobody doubts the efforts, not only of well-known scholars of anti-economism, that went into criticizing the misleading illusion of welfare induced by a deceiving statistics at whose centre lies the GDP or the GNP. Starting from the need to remove from the structure of GDP those activities with "a high degree of incivility" indicated by Bertrand de Jouvenel (1972), through the need to account for the free-of-charge services that are not traded on the free market, proposed by Galbraith (Galbraith, 1975) and all the way to the efforts of scholars that attempt to build a foundation for ecological sciences that interpret the results of economic activity from an environmental perspective, we witness just as many moments when statistics is invited to take an honesty test. It is an exercise involving the accurate representation of GDP and the compatibility between its growth and that of social wellbeing. It also involves the correlation of the "emphasis" on growth with progress in the distribution of incomes, the exploration of the effects of the underground economy on macroeconomic results and the securement of comparative compatibilization of results among countries that use various currencies and exchange rates, etc. If such intentions are to give development meaning and define its ultimate goal—human development—which is nothing less than the inner sense of sustainability, they can only be stimulated and supported.

As an example, we think that it is worth encouraging the attempts, a lot of which are successful and have received institutional validation, of tying the GDP to the pros and cons that production processes inflict on the environment. We refer to the admirable effort to demonstrate that "[e]cosystems are capital assets" (Dasgupta, 2008, p. 3). The solidly reasoned motivation of the Stiglitz-Sen-Fitoussi Report (Stiglitz et al., 2009) is part of the same train of thought. It is worth supporting the type of analyses that show that, calculated in standard manner, GDP does not reveal inequalities or their influence in the distribution of incomes and it does not make the connection between current wellbeing and its relationship to the future. GDP calculation does not consider the stock of natural, social or human capital, the environment, the quality of life and collective services.

Another vice of GDP is less theorized. It is less visible, “painless”, but at the same time misleading and confusing. The ability to use three determination modalities (the output, expenditure and income methods), statistics can elegantly deceive by painting the results according to the interest of those that commanded it. In short, GDP can grow alongside poverty, or conversely and strangely, living standards may appear to be higher than the GDP indicates. The second situation relates to the famous growth through consumption. It is growth fostered by a GDP increase that is not backed by an essential contribution of added value nor by rapidly growing businesses. Wages are the ones that grow and, alongside them, so does consumption. It is less important that behind such salary increases there are no comparable increases in productivity. What matters is the establishment of the paradigm of demand stimulation, bearing strong potential for populism, and which fits perfectly with the ideal of reducing disparities at any price and the securement of distributive justice. Not even Malthus or Keynes, who were drawn to the “mystery of actual demand”, would resonate with this phenomenon. They thought it would be a short-term solution to temporarily balance an offer that remained answered, and in no way as a permanent one. However, the populist appeal of this method explains everything. The innermost perspective of such a mechanism looks like a fan of wages and taxes on wages. The growth of wages is marked as added value. From the increase of added value, one can also raise pensions, which amounts to another “new value”. Taxes on wages raised artificially are becoming more important compared to the real contributions of productive fields to the creation of added value. Added as such to the GDP, they are feeding the most consistent rubric of governmental expenses: “welfare”. This is the main feature of a welfare budget: the collection of taxes and expenses on social security. With the help of statistics, the government may enthusiastically announce a significant growth of the GDP per capita. Time needs to pass until every citizen realizes what he is left with after paying all the taxes—after already having “sold” their votes to the politician with the most generous pledges. It is too late to understand that what they have been given with one hand by the government cannot take the form of a “wage fund” in the Smithian sense, of which they can both consume and invest. Actually, what they have been given with one hand is being taken away with "n" other hands, thus being denied their position as potential investors by a confiscating state that is generally "graceful" with everybody! As long as the voting machine operates with this fuel, it is difficult to reconsider the paradigm. This re-evaluation is necessary and the politicians who institutionalized this machine know it all too well. There are signs pointing in this direction: a lot of money on the market compared to the modest increase in the volume of goods and imports, the deficit of the current account and pressure on the exchange rate. Yet all of this unfolds against the deceiving background of a false growth. By the time the truth emerges, the satisfied people forget that they were deceived and vote again. And so, short-term happiness can be a painful tribute or a deadweight on the foundation of long-term sustainability.

This briefly contoured context highlights the notion of the decisive role of the industry (in the general sense of manufacturing activity) in the process of creating a healthy GDP. Consumption is hardly the source of long-term sustainability, we would say rather tautologically. Industry, with its strong and well-known driving effect, offers support and confers resilience to sustainability. It sanctions the populist immediacy of officially encouraged consumerism. It brings investments and jobs with a sound and real basis and with long-term social advantages. Germany or China insists and persists in maintaining a solid manufacturing base because they learnt that this is the solution to avoid the catastrophic costs of crises. The big and highly performing industry accompanied by the necessarily proven productivism confers the ability to efficiently use raw materials; it facilitates the creation of jobs in directly productive fields with beneficial effects on the development of services and the creation of recuperative and non-polluting technologies. This path, rather than the populist and levelling growth of unsupported consumption, is our only chance to maintain a healthy environment.

The possible impoverishment through growth is another unexplored "virtue" of the GDP. The notions that work does not necessarily bring the corresponding wellbeing and that the GINI index grows alongside the GDP increase are not infrequent. We do not refer to the situation when social polarization is supported by the governing party that, in its search for votes, forgets about entrepreneurs and gives money away to the poor. Rather, we are talking about the adventure towards sustainability of such emergent countries that either try to reindustrialize or start over on this path and connect in a losing manner to the great globalization. It is the situation of those people who, under the social pressure of a need for jobs, consider foreign investments as a miraculous solution, without firm contractual conditions regarding either labour conditions or the environment, just because necessity demands it. The internal anatomy of such a phenomenon is easy to decipher, but politically undisclosable. Concretely, in such countries, the turnover related to foreign capitals exceeds 60–70% of GDP, but the added value lies under 40–30% of GDP. The country "grows", foreign capital comes in with new technology and creates jobs, but pays them "competitive" wages. In other terms, foreign capital "passes through" an economic entity, leaves low wages in its wake and returns home with large profits. The supply creates a demand corresponding to the minimum amount permitted by the laws of competition, which leaves Say dissatisfied, but in full accord with the "naturalness" of the virtues of globalization and with the alleged conformity of the free market. It is growth, yet it is fostered by consumption and the buying capacity of low wages paid by foreign capital, but also helped by the domestic, more "national" wages paid by the domestic capital, which actually make up for the approximately 60% difference in added value.

Within the anatomy and philosophy of sustainable development, the denial of foreign capital is not genteel. And actually, it should not be. At the same time, we believe that the analysis of the often-insidious manner of foreign investments affecting, with certain plusses and minuses, as an algebraic sum, the sustainability of emergent countries should enjoy the same treatment as the issues concerning social and environmental matters. And, concerning the environment and the way its exploitation is related to the GDP, the plea of those who consider the pros and cons of economic growth on the environment and capital, while failing to wonder where resources come from is, theoretically speaking, neutral and "quiet". In a globalized world, both growth and environment belong to the world as a whole. Environmental problems affect us all and we all should be concerned about them. It is politically correct. Less fair is the internal structure revealed by the geography of the phenomenon. While the chlorophyll-penned champions of the New North rest comfortably after their countries impose the interdiction of cutting their domestic forests, the countries forced to contemplate their own forests being cut, often abusively, by companies that cannot do the same at "home" and, unable to do anything about it, cannot rest peacefully at night. It is expedient to export furniture without cutting one tree from your own forest; it's also very sustainable! It is also rather convenient to legally export all the garbage—billions of cubic metres of waste—generated by your industries. However, can you then "compensate" the minus in sustainability generated for other countries with a hypocritical discourse, steeping the subject in the waters of globalization? This is a matter to reflect upon for all defendant sustainability scholars, including the ardent supporters of degrowth. It covers both the exporting practice of polluting industries and the "ecological care" of conserving their own resources and clean environment to the detriment of other countries' resource funds. This is an old practice. It goes on while complying with institutionalized norms and practices, and also exploiting such opportunities as are offered by "cheap", easily corruptible governments. Resources are unique, as the promoters of the philosophy of degrowth claim. Wouldn't a "tax on the opportunity" of exploiting them, wherever they are, together with the prohibition of the immoral unidirectional waste trade, better befit the logical scheme of said philosophy? Apart from royalties and other taxes! This is another subject to think about. One that would provide the opportunity for an update of anti-economism, beyond the invasive and corrosive discourse anti-production and anti-profit. Calculating GDP in the mirror, by considering not only what one has produced, but also what one has taken from and given to others, especially in matters of environment and resources, would constitute a chance for a real gain in honesty, not just regarding the supposedly disinterested statistical manner in which such subjects are treated.

4.2.4 Who Criticized and Still Critiques Profit?

If, on the route towards a "green" as possible GDP and as congruous as possible with human efforts, anti-economism may be saluted, the same thing cannot be said about its allegations against neoclassicism and its current followers, claiming that they deprive economic dynamics of its "soul" and they totally subordinate it to the logic of profit. As we have said, one of the big confrontations of sustainable development is in trying to persuade that profit is not everything, that it remains an important piece of the totality of factors that shape the dynamics of a society, yet this should not mean that the logic of profit must take over people's thoughts and facts or that efficiency is bound to make the very results of human action prone to mercantilism. We wonder, however, did the classicists and the neoclassicists really endorse such a view? Did they only think about profit in and of itself and forget about the social finality of production? As far as we know, the answer is no.

There is no need to catalogue the circumstances in which the analyses of the founders demonstrate that they did not contemplate profit as the end of the road. At least until Milton Friedman, economists did not dissociate profit from the social motivations of the company. The examples that concur in this respect are plentiful. Walras and Pareto, the representative "mathematicians" from Lausanne, did not neglect social aspects. *Studies in Social Economics*, recommends Walras (2010) as a scholar preoccupied with the issue of inequality and distribution. Pareto is a similar case. The dosage of social elements within his works is more than generous. Marshall and Pigou, other famous neoclassical names, do not pose as enemies of social concerns and environment. The former literally writes that "the inequalities of wealth, though less than they are often represented to be, are a serious flaw in our economic organization. Any diminution of them which can be attained by means that would not sap the springs of free initiative and strength of character, and would not therefore materially check the growth of the national dividend, would seem to be a clear social gain" (Marshall, 2013, p. 594). Thinking that education is a chance for the improvement of man and writing that "to this end public money must flow freely" (Marshall, 2013, p. 597), Marshall advanced the idea of a meritocratic perspective continued by A.C. Pigou (1920). The latter also dealt with the idea of income distribution, transfers and social security. To adequately manage state subventions, he thought that it "implies detention under control without excessive leave of absence" (Pigou, 1920, p. 768). Should we add that Pigou was also interested in the moral duty of the employer towards his employees? In other words, the first generation of neoclassical economists was concerned with the matrix of wealth in full connection to its distribution. People at the base of the social pyramid were not excluded from this structure. The structure itself could not marginalize them. Its logic linked the wealth of industry captains to the consumption of the goods they manufactured. If the profit, or at least a part of it, was not invested in activities that responded to the needs of consumers and if those consumers were not employed and given a wage, everything would fall apart. Say was right to explain this objective interdependence, and the neoclassicists agreed.

The Great Depression of the 1930s disturbed the balance between the logic of the profit and society. Consequently, its reinforcement required additional efforts. Keynes took over this task. Preoccupied with the explosive potential of inequalities in a time of crisis, he writes that "for my own part, I believe that there is social and psychological justification for significant inequalities of incomes and wealth, but not for such large disparities as exist to-day. There are valuable human activities which require the motive of money-making and the environment of private wealth-ownership for their full fruition. Moreover, dangerous human proclivities can be canalised into comparatively harmless channels by the existence of opportunities for money-making and private wealth, which if they cannot be satisfied in this way, may find their outlet in cruelty, the reckless pursuit of personal power and authority, and other forms of self-aggrandisement. It is better that a man should tyrannise over his bank balance than over his fellow-citizens" (Keynes, 2018, pp. 332–333). The other part of the barricade does not escape his scrutiny. The rentier appears to him as a profiteer, a hideous and egotistical character. What he does, namely the accumulation of money for its own sake, is illogical; it is an affront against the idea of putting people to work in a time when turmoil took away their jobs. Thus, accumulation for the sake of accumulation is unacceptable to him, especially in troubled times. He convincingly writes that, after the issues will be solved, "the love of money as a possession -as distinguished from the love of money as a means to the enjoyments and realities of life -will be recognised for what it is, a somewhat disgusting morbidity, one of those semicriminal, semi-pathological propensities which one hands over with a shudder to the specialists in mental disease" (Keynes, 1963, p. 369). Thinking about the suffering of the many, Keynes allows himself to perform a sacrilege. He calls for the state to help the market. He calls for its involvement in the fields of redistribution and monetary relaxation. His call was motivated by need and the solution was thought to be short term. The unintended costs of his initiative amounted to two things: (a) the state became and remained a huge redistribution machine; (b) the public was caught by the pandemic of revendication-manic—it realized that it can profit of the fact that it deserves compensation for the failings of the market. Theorizing on these two levels has granted Keynes a chance to popularity and a place in heaven.

The first effort belongs to Paul Samuelson (1955), the author of *the neoclassical synthesis*. To him, market infallibility is questionable. He also has a problem with the logic of profit. In a word, by extending the fissure opened by Keynes, Samuelson sees redistribution as the buffer solution to the inadequacies occurring in an unsupervised market that is guided exclusively by the pursuit of profit. The provoking opposing position of Milton Friedman on this matter has fed, once again, the ascending trend emphasizing redistribution and social justice. Considering that acting in the interest of society is an irrational strategy and that the only objective of the corporation is to yield the largest possible profit for its shareholders (Friedman, 2007), Friedman deeply plants the seed of a polemic whose ramifications have reached our times. After being provided with a path and a motivation, the philosophy of redistribution and the inequality motif would become "hardcore" matters of scholarly works, some of which would receive the highest scientific consecration. In typical manner, Paul Krugman and Joseph Stiglitz treat the reduction of income inequalities, the regulation of markets and state interventionism as favourite subjects. Even the coarse perspective of treating the state budget as a bottomless pocket is appropriated for the sake of social peace. Inequality is ugly; it has a price and this price must be paid by those people wearing blinkers, who cannot see anything beyond profit. The latter are reprimanded by Stiglitz, who writes that "those in the business sector see things differently: without the restraints, they see increases in profits. They think not of the broad, and often long-term, social and economic consequences, but of their narrower, short-term self-interest, the profits that they might garner now" (Stiglitz, 2012, p. 112). Regulation, redistribution and reduction of inequalities seem to be the keywords that define the new paradigm of sustainability in the eyes of the post-Keynesian super-qualified scholars of anti-economism. Individualism and the perturbing mechanism of profit are being left out. Regarding economic approach: "...this individual-centered, bottom-line economics, tailor-made for America's short-term financial markets, is undermining trust and loyalty in our economy", is what Stiglitz strongly believes (Stiglitz, 2012, p. 104). With such a bridgehead, certified by famous signatures, the road is paved for redistributive and justice-driven anti-economism.

A series of studies provide support to viewpoints that consider a fall in inequality prompted by state intervention would do a lot of good not only to "liberal" Krugman's America (Krugman, 2007), but to the entire planet. Thus, Roland Benabou (1996) claims that there is no conclusive evidence to lend credibility to those scholars who state that redistribution or the constraint of credit would negatively impact economic growth. On the contrary, as this author thinks and exemplifies, South Korea got ahead of the Philippines with respect to economic performances precisely by using the tool of redistribution. Following the same train of thought, Torsten Persson and Guido Tabellini (1991, 1992) think that inequality hampers the economic growth of democracies, and the concentration of real estate ownership worsens the phenomenon (Persson & Tabellini, 1992); in addition, inequality generates policies that infringe upon the legal estate, generating dysfunctions in the domain of appropriation of incomes generated by investments (Persson & Tabellini, 1991).

4.3 From Anti-Economism to the Court of Distributive Justice

4.3.1 From Arithmomorphism Towards Dialectics and Distributive Justice

What are the explanations for anti-economism becoming the updated version of economic philosophy, so strongly present nowadays? We find out from inside sources that the neoclassical line with its dry abstractionism and its exaggerated quantitative orientation were to blame. This is allegedly a response intended as a return to an economics version resembling a social construct that can be used to treat and explain not only the dynamics of the wealth flow, but also its distribution. Using arithmomorphic models, this is not possible. Some dialectics is also needed. We are invited to believe that anti-economism takes aim against a method, too abstract and unsuited for sensitive subjects. We may be allowed to say that we are in the domain of outward appearances. In fact, we can see another thing too. We can see how, keeping an eye on inequalities, anti-economism covers a much wider front. The range of criticisms include accumulation and production, as well as profit and productivism. Nothing that comes from the past and supports the logical structure of economic dynamics escapes. This critical analysis is directed against fundamental ideas originating not only from the neoclassicists, but also, or mainly, from the classicists. The criticism of the mathematized model provides only the opportunity; the real target is somewhere else.

It is not only anti-economists that criticize the modelling excesses of economics. The controversies on this subject are old and haven't been laid to rest yet. It is rather interesting that an anti-growth economist, a pioneer in the field, did not shy away from mathematics in his economic analyses. We are talking about Nicolae Georgescu-Roegen. Convinced that mathematics may be really useful for the economist, he wrote that "of all men of science, economists should not let their slip show by opposing the use of the mathematical tool in economic analysis, for this amounts to running counter to the principle of maximum efficiency" (Georgescu-Roegen, 1971, p. 331). From him and other economists, of which the Austrians are the most persuasive, we also know that not everything can be subject to mathematics in economics. Where possible, the mathematical route shortens processes and increases efficiency; where this is not possible, qualitative analysis is called to help. The problem is one of proportionality. The exact measure is not set by anybody; the context fixes its amplitude. Declaratively, anti-economism has a problem with exaggeration. Actually, in other areas as well, it does not claim to go all the way with its criticism. It is not accumulation and profit, but rather accumulation for the sake of accumulation and profit for the sake of profit that seem to be the deficiencies. They just seem, because if we visit their home ground and observe their obsessive reflex of putting distribution before production and making social justice their basic preoccupation, we are constrained to see that productivism and profit per se are their actual enemies.

If the allegation of quantitative extravagances is justified, the reproach in bulk and the association of this analytical propension to the neoclassicists is unfair. As we have shown, the alleged lack of humanity and the preaching of cold rationality is not the summarizing feature of the neoclassicists. All of them, or almost all, have ended their scholarly journey not in the ivory tower, but in the social agora. It is true that between their abstract works and the ones in which they tackle the issues of inequality and distribution, a long period of reflection, decantation and doctrinal clarification was needed for some of them (Walras is a typical example—22 years have passed between the publication of *Elements of pure economics or the theory of social wealth* (1874/1954) and *Studies in Social Economics* (1896/2010)). However, this cannot elucidate the attitude of dissociation between the technical, marginalist analysis of economic mechanisms and the social analysis performed by the same author. We want to say that even though the work concerning social issues was produced as a late consideration or an annex to an already finished construct, this does not justify its relegation to the footnote category, or not even there. However, this is how things happened. Every time Walras and Pareto are mentioned, such references are made predominantly to their mathematical works, their "major work"! Smith's (1977, 2004) case is no less conclusive. His *Theory of Moral Sentiments* is, in regard to the statistics of citations, a pariah in comparison with the *Wealth of Nations*, given the fact that the former introduces us to the universalism of a remarkable classical mind.

We do not know why things happen like this. What we do know, however, is that we perceive a shadow of diversion seeping into our analytical case. We severely criticize profit, production and productivism by accusing a method, in order to do what? To venture, subtly yet pedantically equipped and protected by the label of science, to places that belong to uncaused causes, towards the initial theft of property, the violation of natural equality, the immoral establishment that prefaced the new world whose name is uttered with more and more animosity—capitalism. We need not quote Proudhon anymore. It suffices to quote Marx if I do not hide my belonging to an ideology and my name is Gorz, Haribey or Piketty. Because Marx, actually, with his vocation for synthesis, criticized everything that could be criticized related to capitalism. Nowadays, we do not have to do this anymore; nowadays, we have to go beyond him. Or we may stay in his "cathedral", but at the same time build a "political economy of de-commodification", criticizing and getting rid of not only economic liberalism, but also of "a certain Marxism" in order to loyally return to the great apostle (Harribey, 2010). In addition, what Marx did was to enable the effort of stultifying whatever came before him that could have been deemed as revolutionary. The law of accumulation was a revolution; revolutionary was the role of the entrepreneur for both Smith and Say; revolutionary was the hedonistic principle inherited from Quesnay; specialization doubled by cooperation was included in the same category. And finally, the social order based on the dynamics of unequal participants in the act of production, with opportunities of changing places according to performances, provided completeness and a meaning to the landscape. And what did Marx do? He broke the mechanism. According to him, the law of accumulation became the law of capitalist accumulation: from surplus value, capital takes more and more, while the worker gets less and less. Capital itself becomes "dual"; what matters is the core, the exploitation relation it conceals. The entrepreneur turns into the satiated bourgeois with a stomach whose dimensions poor Smith could not have predicted when he dared to think about "the invisible hand". The market turns into a plan and profit is plagued—the ways of appropriating it are interpreted as forms of exploitation. Inequality? Let's be serious!

We are not entitled to say that the promoters of anti-economism follow in Marx's footsteps. Everyone walks their own path. We are only interested to find out whether the direction they suggest, given its strong representation, supports economic sustainability or not. As we have hinted from the very first lines of this chapter, we have serious doubts believing that such a path leads to an increase in sustainability or resilience. In support of this hypothesis, we shall exploit at least two hypostases: distribution before production; the anathema on profit.

4.3.2 *Distribution Before Production. The Illusion of Distributive Justice*

We would not insist on such an issue if its delineation was not linked to illustrious scholars. Nobody believes that they might be able to allocate something not yet produced. One gives what one already has, not what one will produce in the future. Saying otherwise would risk raising suspicions. However, this is not the case when you have on your side the authority of the idea, nourished by the consecration that academia has bestowed upon you. An example in this respect is the Stiglitz-Sen-Fitoussi Report (published in 2008–2009). It starts with the emphasis on the well-known limits of GDP and the exclusion from its structure of things that are hard to include: inequalities in distribution, the quality of air, water, soil, etc. There is nothing bad or new in the authors' objectives, in their proposals of adjacent indicators or in their preoccupation of ascertaining how sustainable is current wellbeing for it to be transmitted to future generations as well. What is worth remembering, however, for the purposes of our analysis is that the five members of the Commission chaired by Stiglitz, Nobel Prize laureate for economics, instead of analysing why GDP is not increasing all over the world, they focus on the reasons why it is unequally distributed. Economic performance is not seen with good eyes. It seems that distribution, health, the quality of the environment and transferable capital stock were more important for them, although all these depend on economic growth. The Report's list of recommendations begins with "Recommendation 1: When evaluating material well-being, look at income and consumption rather than production" (Stiglitz et al., 2009, p. 12). This idea is delineated more precisely in Recommendation Four: "Give more prominence to the distribution of income, consumption and wealth" (Stiglitz et al., 2009, p. 13). The role of economic growth as a starting point and orientation for all the other recommendations does not occur. In no place in the text is there any reference to the way in which GDP is compromised and transmits false values. The crisis, which started in 2008, the same year when the Stiglitz-Sen-Fitoussi Commission was assembled, sanctioned precisely this exercise, of "generating" false values through credit granted out of thin air.

The Report was prepared at the order of the French presidency of the time. Those who paid the bill were interested in creating a tool of political marketing; the scientific truth, in this case, would not have served them. The ones who implemented the idea did not have serious issues with the "interest of consciousness" as James Buchanan (Tollison & Buchanan, 1972) would say. If the moment has passed, something else has not passed and indeed does not seem to be going away, namely the philosophy of the report. This line of inquiry has been fostered by previous declarations, such as the ones of Rio and Johannesburg. In neither of these are creation and production put at the forefront. The famous Millennium Declaration adopted by the UN in 2000 made no exception. From the content of the resolution, it is only at point 20 that we find a mention on production and economic growth: "We also resolve: to develop and implement strategies that give young people everywhere a real chance to find decent and productive work" (UN General Assembly, 2000, p. 5).

The road to heaven is paved with good intentions! But it is hard to understand how one plans to get rid of poverty and starvation, provide good quality education, clean water and health care, achieve equality between genders, etc., without placing growth first. How can one live well in the long run if the first place in the menu of sustainable development is obstinately claimed by distribution? It's true, one may live well in the short run, deceived by politicians in search of votes by the offer of a job. Such jobs would be created, as Minsky explains, "directly", out of thin air. It may be a false job, possibly in the "services" sector, that disappears on the first occasion when the economy is hit by a crisis. Then, when the economy will no longer be able to afford such luxuries, the purge would start precisely from the service sector. When such a construct is endorsed by famous scholars, its design is not called into question. Yet with all the risks, questions must be asked and people must be shown and persuaded that the sources of a good life are, like it or not, work and production!

Another institutionalized response drew attention! It belongs to a group of German and French experts in economics. A Report of the Economic Analysis Council and the German Council of Economic

Experts formed a reply to the Stiglitz-Sen-Fitoussi Report. This reply puts causal relationships on their natural course; rising performance appears as a prerequisite for all others. The reply suggests that the version of sustainability proposed by the Stiglitz-Sen-Fitoussi Report should be amended to also include issues of economic sustainability and not focus exclusively on environmental concerns. Hope is pinned on the growth of GDP to reduce disparities and poverty. The essential idea of this reply is the following: without economic and financial sustainability, the whole system suffocates and there is no longer any reason or time to talk about the quality of life.

This response is sound, yet singular and very much against a widely promoted trend, thus having few chances to coagulate a solid critical mass. The pronounced justice-oriented nuance works in the same direction. Besides the lack of consideration for the major interests of developing countries (Somalians have nothing to redistribute, otherwise they would!), the Stiglitz-Sen-Fitoussi Report is also only vaguely connected to the hypotheses of the founders. Smith, Ricardo and Mill would have had something to say on the subject. Even a biased understanding of Ricardo—an economist of distribution par excellence—would have brought extra solemnity to the text. Smith (2004) would have been all the more clarifying. In fact, redistribution finds a logical and natural foundation in the Smithian example of the rich man's stomach, which is too small to accommodate all the cravings of its owner. It is logical, fair and natural to distribute that which one possesses in abundance and does not use himself, thus fulfilling the needs of others and also working towards the general interests, an idea that is explicitly present among the "moral sentiments" that preoccupied the great classic. If the stomach example is rather abstract and cold, his pieces of advice presented in his *Theory of moral sentiments* are actually comforting when he talks about fame, happiness, money, wealth, decency, tolerance, justice or virtue. From this book, we learn about the propensity to not stop and to be endemically dissatisfied with what we have and the wish to always have more. Smith outlines here two great paths towards happiness: the path of wealth and the path of wisdom and virtue. Even though the former is visibly more attractive, it is not the most sustainable, as Smith tries to persuade us. And he succeeds by not placing on the scales of happiness the fame of wealth, the eulogy of material possessions, intolerance and envy. No, the equation of happiness includes health, the lack of debts and a clean conscience. Russ Roberts (2015) in his splendid book *How Adam Smith can change your life: An unexpected guide to human nature and happiness* portrays this image of the humanist Smith, so overlooked and misjudged in economic literature. In other words, the mechanics of the full stomach that needs to overflow towards other people is backed by Smith's unparalleled philosophy of distribution. Actually, beside the theorization of the role accumulation plays in the production process, the ingenious way in which individual interests are conducive to the satisfaction of general interests if left free, constrained by nothing but the Law and connected to the "great chess board of humankind", represents one of the greatest accomplishments of economic science.

The scholars who positioned themselves as followers of the doctrine of the great economist, of which Hayek and Friedman are only two happy examples, took full advantage of the naturalness of the Smithian philosophy in matters of distributive justice. Friedman showed how the government faces great difficulties governing, failing to understand anything from Smith's trepidations on whether it was better to contribute to the state or directly to a private school if one wanted to redistribute some of his wealth for the good of education (Friedman, 1962, 1980). Hayek (1976) made exemplary efforts to elucidate the mirage of social justice. He came to the conclusion that *The constitution of liberty* (Hayek, 1960) can offer the legislative framework admissible for the wished-for equality of opportunity. However, the fact is that every person has his own natural endowment and we come unequally into this world, before this Constitution takes us over which ruins the scheme of achieving equality of opportunity. Hayek persuades us that "[i]n a continuous process this initial position of any person will always be a result of preceding phases, and therefore be as much an undesigned fact and dependent on chance as the future development" (Hayek, 1982, p. 130). What are we left with? We can only admit that, with respect to the distribution of results among participants, "[t]he rationale of the economic game in which only the conduct of the players but not the result can bejust" (Hayek, 1982, p.70). And we are also left with Friedman's warning that "[a] society that puts equality — in the sense of equality of outcome — ahead of freedom will end up with neither equality nor freedom" (Friedman, 1980, p. 148).

It was not easy for Smith and his followers to indicate with certainty what is right in a distribution of wealth. Many of his questions remain unanswered to this day. What is known, however, with certainty, is that he never renounced one principle: the absolute importance of production and growth. That is, we first produce and only then can we talk about finding happiness. And on our path to this end, we will never be equal. Only inequality is natural, and thus, it is the initiator of creation; equality isn't. Unless constructed and ordered, there is no equality. Remains of primitive naturalism or of the popular dream close to anti-economism may create illusions. Such illusions do subjugate, as distributive equality is an "all-too-human" dream! At the same time, such illusions mislead inasmuch as we do not possess or we claim not to possess a clear answer to the question whether the egalitarian ideal as a working hypothesis can be fulfilled with less or more growth. There are studies that try to give credit to the notion that inequality hinders growth and vice versa (Benabou, 1996; Persson & Tabellini, 1991, 1992). The history of economic dynamics shows otherwise. Putting things in historical perspective, it illustrates that the pathos of equality condemned growth. Communism is the perfect example, suitable to serve as "literature review", indicating without any doubt that this is the case. And today, the "Chinese transformation of Europe"—a threat which was signalized by both Tocqueville and Mill—bears the same message, but from another perspective. Red China has been considered the cradle of distributive equality par excellence. However, today's China paradoxically becomes a place of inequalities. The billionaires of a spectacularly growing China defy any notion of equality as a source of growth. It is the example of a country that understood that, through production and growth, it could solve its many and serious social problems. Admittedly, that came at the cost of growing inequality; yet it has given China the chance of not needing to compare itself but to itself to decide whether the price is worth paying. And it does pay it, understanding at an exceptional scale that it can no longer survive emulating the utopian construct of egalitarianism in poverty and that the process of creation, of the great production, comes at the cost of higher inequality; that forced equality steeped it into poverty and that it is no longer worth suffocating its production with ideologies. Does China take into consideration the necessary and permanent redistribution of goods to resonate with the correction mentioned by J. S. Mill to constantly secure equality at the starting point? How could it do it? Maybe Malthus would be a better counsellor in its novel planetary endeavour. How long could an overpopulated China bear inequalities? However, Malthus set some milestones in this regard, and every nation should choose its line of conduct accordingly. He warned that "redistribution from rich to poor should not be carried too far. An excessive redistribution from rich to poor would result in slower growth or depression by reducing the motivation for investment and by reducing savings and the supply of capital" (Pullen, 2016, p. 44).

In conclusion, redistribution is not a process in itself or an independent piece on Smith's chess board. It belongs to a causal chain, being influenced by and influencing other elements in its turn. Its promotion while forgetting the other components of this engine is only justified in one case. It can be accepted that redistribution is more important if and only if we assume that production has no issues and works by itself. Would this be the assumption of scholars who place it before production?! It's hard to believe! Does production work by itself in Kenya, the poor country of the experiments of the latest Nobel Prize winners—Banerjee, Duflo and Kremer? (Duflo et al., 2008, 2011) Their work has the audacity of those that go against the current, though they were not the first. The pioneers with same-level acknowledgements were Daniel Kahneman and Vernon Smith. Despite J. S. Mill's warning that abstract speculation rather than experiment is the strong core of economic science, they, just like the latest Nobel Prize winners, tried to achieve the contrary. They tried and seemingly succeeded to validate the heretical Sismondi. It is the latter, rather than the classicists, who found that the true object of economic science is human condition, man, his life and the institutions' impact on him (de Sismondi, 1838). The humans in Kenya, the study of their life, their small questions, the identification of the poorest categories entitled to social security, the rate of use of fertilizers by the farmer, the effect on school attendance of programmes of intestinal worms' elimination compared to programmes of employment of additional teachers—this is the workshop of scientific experiments in which Banerjee, Duflo and Kremer worked. The emphasis on experience as the main source of information can only do good to economic science. However, the real issue is different. The observation of facts must form the foundation of the theoretical edifice. M. Blaug (1985) convincingly demonstrates it, and not only him. But where is the theoretical edifice? What is the theoretical basis or the theory that informs the empiricism of the mentioned authors? Isn't the trendy matter of poverty eradication part of a theory? It is, we think, and this is the theory of growth, which is sustainable and can help to causally explain that the children of Kenya have health issues and the country has many poor people because its economy does not grow. This is the starting point. It is striking to find out that the elimination of intestinal worms is more efficient than the supplementation of teachers. But how does one link this finding to the general plan of a theory intended to include all poor people and emergent countries? Focusing on small and real issues may transform the economist into a "plumber" or an "architect" to show people "how it's done"; it may help solve some urgent social problems. But if in this endeavour, people forget the big picture, they have no chance to inspire a healthy, resilient and long-term behaviour. Kenyans must be taught how to "fish", how to start growth, maintain it and solve their social difficulties using their own forces. They must be made aware that without growth, there is no food, they cannot be healthy and they lack the means to buy medication when they get sick. Mistaking the symptoms of the disease for its causes, one cannot make gains in regard to sustainability. The only ones to gain are external programmes that gather ever more poor dependencies. Hence, the Brundtland message is deleted!

4.3.3 The Poison of Profit and Productivism

Profit and the entrepreneur are the soul of capitalism to the same extent as masses and redistribution define the soul of communism (socialism). The former case provides benchmarks for the necessary economism. The latter one provides the breeding grounds for anti-economism. Profit belongs to the newly created value, alongside wage. From the perspective of the objective theory of value, the two conceptions are conflicting. Marx was highly drawn to this subject and condemned it to eternal discord. The neoclassicists did not escape the harmful and conflicting scheme of a zero sum game. But, unlike Marx, the little or big bourgeois did not exit the scene. He became the stuffed wealthy person who eats too much, so nothing is solved by the fact that planet Earth has enough food for everyone! Harribey's "bourgeois imaginary" remains an operating category that is present in the landscape that should lead to Andre Gorz's (2008) *Wealth without value, value without wealth*, where work rather than profit-bearing capital is the first violin. The bourgeois, animated by the wish to obtain as much profit as possible to the detriment of the employee, is still criticized by all scholars who find support in Marx and Marxism. The bourgeois cannot befriend the worker as the latter, being unfairly exploited, denies the former's friendship. They do not belong to the same class! The former offers lunch. The latter is not willing to pay for it! He rather looks towards the state and the government, from where he expects a "well-deserved" gratuity (Marinescu et al., 2012). He knows that the government creates neither wealth nor jobs. But it can redistribute them. He prefers to pay taxes and duties than to admit to any debt towards his employer (McGee, 2009). It is in vain that Friedman utters the warning that "There is no free lunch!" The masses consider themselves entitled to this gratuity; they want the lunch, and afterwards, they can be passionate about the big promises of the government. They even want equality without the entrepreneur.

This is, in a highly abbreviated form, the essence of the anti-economism that targets the entrepreneur and the profit. Its aim is, unfortunately, not that far away from the one set by Marx and his followers. In Marx's work, as well as in Mill's, albeit in incipient form, the morality of profit is starting to be questioned. For them, there is an uncaused cause in this regard; there is a theft! That was the starting point. Down this path, capital that has been accumulated and invested is still the worker's unpaid or low-paid work. There is no possible ethics to account for the robbery. This process must cease in order to put an end to such lawlessness, and with it, to accumulation, profit and the profiteer. What anti-economism wants is, more or less, the elimination of the bourgeoisie and profit.

Brundtland's message also claims that profit is not everything; there are so many other important and already present aspects in the equation of life. The logic of profit should not subdue and transform us into robots animated by the soul of an accountant.

It is true: profit is not everything. It is certainly not everything for people that go beyond the daily hedonistic calculation, for artists, writers, teachers, lawyers, doctors, priests, etc., even though we have some doubts on this point, as well. But what about the entrepreneur? If profit were not important to him, he would be called a dreamer, a star seeker. Friedman was accused of picturing the entrepreneur in such a manner. But can we really reproach the entrepreneur that his social agenda is less important than his concern for profit? We may congratulate him if he does have a social agenda, but we cannot enforce one on him. Thus, we believe that the assertion according to which life should not be reduced to profit must be construed at the crossroads of two dimensions: (a) one of the entrepreneurs; (b) another of the society as a whole, of which, admittedly, the entrepreneurs are part of. Irrespective of our perspective on such things, profit is necessary. The first category cannot do without it. The second has the luxury of pretending to be able to do without it! However, to think of entrepreneurs existing without profits is nonsensical. At the general level of the masses, we may understand the hypocrisy of morally condemning those of us who are more industrious, bolder and luckier. Envy only disappears under absolute communism. In a free society, it obsessively follows us everywhere and makes us consume a large quantity of energy that would rather be used more efficiently for other purposes. This is as well the creed of anti-economism, shared by all its supporters who believe we can also live without profit. However, we believe nothing is done in its absence. Even charities cannot operate without the money that has been earned "elsewhere", in the search for profit. Does the Church operate without money? The Vatican has its own bank. Money flows in from profitable areas, notwithstanding Christ's ancient anathema on merchants. A new idea does not become an applied technical innovation without a knowledgeable entrepreneur interested to gain profits out of its implementation. Science and knowledge need innovations. Nobody transposes an idea into practice without thinking of profit. Like it or not, science does not work without money either, and without science any project for keeping under control the ecological footprint is nothing but an illusion.

The simplicity with which profit was included in the configurative sentences of the growth paradigm has made them prone to being treated as simple, commonplace truisms. Who is now willing to leave aside their main concerns to ponder on the importance of accumulating a part of the gained profit as a really revolutionary hypothesis? Against such a background, emptied by forgetfulness or conscious evasion of the fundamental sources of economic dynamics, profit and productivism become the main targets of the criticisms of all anti-economism supporters. And again, the classicists and neoclassicists were accused of all evils. It is often forgotten that the classicists were not the first scholars who saw the benefits of profit. Before them, the mercantilists saw a certain opportunity for progress in the enrichment of the nation through the acceptance of interests and profits. Freed from the guidance of the Church, their doctrine considered the wishes for gain of the individual citizen as something natural and legitimate, achievable without having to bear public opprobrium. If we intend to oust all scholars who have thought of profit and wealth as divine blessings, we should start with the mercantilists rather than the classicists or the neoclassicists. In a rather efficient manner, anti-economism reckons with the latter, but scores a victory against mercantilism as well. Because what is the cultivation and the maintenance of sentiments of envy and hatred for the wealthy, especially when wealth is inherited, if not a manifestation against the very philosophy that prefaced the modern world?

Admittedly, Smith, Say or Bastiat share the "blame" of creating the image of a world with profit at its centre. However, they gave us a resilient lesson on this subject, sustainable in all its aspects and frustratingly moral. It is so solid that anti-productivism and the anti-profit dichotomy are like a couple of boulders thrown with a hastily assembled catapult at a construction with a structure too sturdy to falter. What is this about? It is about the fact that all classicists, obstinately Smith, have shown us that the keyword of social progress is growth, without growth life is just an empty word, and even liberty amounts to nothing unless it comes first in economic form. And growth is hardly possible without the impulse given by profit, which generates all enterprising energy. Using the words of a fine and civilized man, Smith tells us that it is not right to spend everything that we have produced. On the contrary, frugality is appropriate. It is adequate that a part of profit is accumulated and reinvested to sustain new growth. Such new growth would bring new real jobs and surplus of goods. It is true that these considerations about the fact that the profit drive must be cultivated and defended appear nowadays as irritatingly common. It's just that this profit drive has changed the face of the world. Economic historians know it all too well. It is not the only "command" passed down by the said classicist. He knows and affirms that the equation of profit is true in an interdependent world, in a network. One cannot earn wealth by oneself and cannot enjoy it alone, in a Robinson Crusoe manner. In a climate of mutual trust (for which money serves as the defining binder), the banker depends on the entrepreneur and vice versa. The equation of profit—industrial, commercial, banking, etc.—connects everybody to everyone. The natural rate fixes the boundaries of their playing ground. The earning of the entrepreneur, wherever it occurs, is not only his. He shares it upstream and downstream with the people to which his business is connected. Earning by means of earnest means and in compliance with the rules is not in conflict with ethics. So, what must be repudiated? Is it the capitalism of Smith and the other classicists, with ethics and morals clearly defined? Is it the capitalism rooted in a free market, comprised of free people in the noblest sense of the word? Is it the ethics by which state abuse and monopolies were restrained and failures were punished by bankruptcy? Does Smith's economy resemble today's economy? No, today's economy resembles the one from the writings of Perroux or Williamson. One where monopolies and deceiving methods are admitted in the pursuit of profit and respect for nature and animals is lacking. Such monopolies have nothing to do with the "ethics of the shoemaker" who wishes to grow his business by reinvested profit. A market with monopolies escapes the Smithian philosophy; it cannot protect the workers. When the monopoly is larger than the state, who can punish it when it evades the law? Forsaken or deceived by the state, the market is exposed to the Smithian "counterweights": thefts, confidence games, corruption and the pursuit of profit in spite of the law. Insensitive to ethical considerations and focused only on balance sheets, the capital of this market is free to remunerate production factors inequitably. This happens because competition was destroyed within the stages of building natural prices. The profit that results from such exercises cannot be fair. In this case, yes, anti-economist does actually serve a purpose. However, in front of such a wide avenue of attack, critics do not think of the representatives of today's capitalism, but of the one from "back then". They fight against the founders even though, if the latter were teleported from the place they are currently inhabiting, they would be embarrassed at the manner in which the "child" in the *Theory of Moral Sentiments* evolved.

We are convinced that with respect to their credo and argumentation, the supporters of anti-economism, those who consider profit-driven growth as a threat, build their convictions on such premises. They think that capitalism is to blame because it came to be using the recipe of the classicists. Nothing could be further from the truth. Instead of wondering whether profit is to blame because it defines the paradigm of development, the supporters of anti-economism would benefit more if they started with an inventory containing the aberrations of today's capitalism compared with the capitalism of the Smithian epoch. They would see, at the same time, that the antidote to today's capitalism, that trivialized the philosophy of profit, is not socialism. Per contra, a return to the founders and their breed of capitalism would be a path that would bring some neatness to economism and its ethics, but also to the social and environmental domains. It is only by envisioning capitalism devised and run according to Smithian canons that we can correctly assess the logic and the ethics of profit. Beyond it, we find the illusion that the right to laziness and stargazing might one day infect the entrepreneur. And when that day comes, profit will no longer determine them to get out of bed to cook us a delicious lunch.

We will never get rid of profit and productivism. The digital era with its great programmers has its own entrepreneurs. Regardless of how he wears his suit and tie, whether he is a Harvard alumni or a hard working labourer teeming with ideas, the entrepreneur of the twenty-first century will not escape the Smithian scheme. He will still have profit in his DNA, without which the drive for creation becomes null. Meanwhile, the philosophy of anti-economism with its anti-productivism and anti-profit mantras proves a pamphlet-like endeavour. It's only usefulness comes in the form of a break to excesses.

4.4 Pikettism—The Zeal of Quantitative Levelling

A version of anti-economism that is gaining ground, intendedly influenced and bearing unsustainable fruit, is Pikettism. We link this production of ideas with a strong hook in regard to social and ideological areas to the name of the economist Thomas Piketty. But it's not just him; his books and articles as well as his media interventions catalyse a state of mind for whose coagulation several other authors cooperate, Anthony B. Atkinson, Emmanuel Saez and Gabriel Zucman (Atkinson et al., 2011; Piketty & Saez, 2003; Piketty et al., 2018). *Capital in the XXI century* (2014) and its extension of 1,232 pages, *Capital and ideology* (2020), assert everything that a socialist can say, and more, about capitalism, inequality, inheritance, property, capital, wealth, accumulation, morality, state, market, models to follow, social justice and the like. They assert such things using the well-known tone, method and customs of anti-economism: the said fundamental categories and institutions of capitalism possess a satanic characteristic. All or almost all of them, this author informs us in his latest opus, "are social and historical constructs, which depend entirely on the legal, fiscal, educational, and political systems that people choose to adopt" (Piketty, 2020, p. 7). What he means to say is that everything we may have studied about market, competition, profit, wages, capital, etc., are not what the classicists, neoclassicists or our teachers said they were. They are actually what Piketty tells us about them. Not even "Inequality is neither economic nor technological; it is ideological and political" (Piketty, 2020, p. 7). From this point of view, we deserve to be shamed just as much as those who believe that inequality may even be natural, because the "Naturalization" of inequalities belongs to a time long gone!

So, what does Piketty bring to the field of anti-economism? The manner in which he builds his argument, his premises, his methodological support and conclusions recommends him, or so do people say, as special and even attractive. See the number of citations of his books. We believe it is the subject of his analyses, an obsession, that determine the destiny of his works. In dealing with the economic results and all its secondary products—pensions, taxes, inequalities, distributive justice, etc.—Piketty's works and Pikettism in general amount to no more than a swan song, a mourning at the deathbed of a world considered crooked and perpetually unequal, which only some Messianic project may help it sort its problems. Focused on passionate subjects such as envy and resentments, boiling up in fury against the wealthy and offering solutions by pointing to a world we are encouraged to dream of, Piketty's works include curiosities that are worthy of our attention.

We believe his bestseller, *Capital in the XXI century*, summarizes his ideas. Thus, we will quote from it and also from the reviews that we consider relevant for our initiative.

A complete and careful reading of the book as well as of the critical observations of the reviewers entitles us to believe that the work should be deprived of its conclusions, as long as it faces big problems when it comes to defining the working concepts and the large empirical (statistical) data it uses. The conflation of wealth and capital, the exclusion of human capital from analysis, the lack of representative data for the domain in question, the hypothesis of a 100% savings-consumption proportion, the lack of a causal link between his big methodological invention—the r > g law—and inequality, etc., are enough elements that disqualify the veracity of his endeavours' results. Deirdre Nansen McCloskey (2014), Lawrence E. Blume and Steven N. Durlauf (2015), Joseph Stiglitz (2015), Francois Facchini and Stephane Couvrier (2015), Jackson T. and Victor P (2016), Brecher R. A. and Gross, T. (2017) emphasize such methodological omissions. For the purpose of our endeavour, we confess we are less interested in the methodological rigor and the accuracy of the data used by him, even though we do know how important they are in defining the scope of a work. We are interested in the theoretical construct, as in this area Piketty pictures himself as a nonconformist, declaring that he is anti-profit, anti-property, anti-market, anti-inheritance, anti-entrepreneur, anti-capital and against all things that might define "destructive" economism! At the same time, we want to see whether the things that he puts in their place have anything to do with sustainability and resilience.

Inequality, with all causes that deem it undesirable, stuffs most of the paradigm of Pikettism. The premise of his judgements on the subject seems civilized. Avoiding "any economic determinism in regard to inequalities of wealth and income" (Piketty, 2014, p.20), we are invited to start things right, with an analysis built on the hypothesis that "[i]nequality is not necessarily bad in itself: the key question is to decide whether it is justified, whether there are reasons for it" (Piketty, 2014, p. 19). The underlying support, the dubitative and labyrinthic way in which the analysis is conducted and, in particular, the proposed final scenario prove that the premise is no more than a delusion. It is disarming to find out that "[a]ttributing a monetary value to the stock of human capital makes sense only in societies where it is actually possible to own other individuals fully and entirely" (Piketty, 2014, p. 163). Deceitful is also the nice promise that the motivation of Piketty's approach is hardly to denounce "inequality or capitalism per se (…) [b]y contrast (he is interested in) (…) the best way to organize society and the most appropriate institutions and policies to achieve a just social order" (Piketty, 2014, p. 31). When we find out that even human capital can be a source of disparity by allowing the shift of patrimony structure through the rise of the middle class or that the model of the former USSR could be imitated due to its coherence and rigor, we come to the desolate understanding of what meaning bears the rule of law and a just society in Piketty's view.

On what points of support does this seemingly scientific and highly erudite endeavour rests upon?

Firstly, on the false assertion that "[t]he economists of the nineteenth century deserve immense credit for placing the distributional question at the heart of economic analysis (and that) It is long since past the time when we should have put the question of inequality back at the center of economic analysis and begun asking questions first raised in the nineteenth century" (Piketty, 2014, p. 16). The main theme of the nineteenth-century economists, their huge merit, was to deliver knowledge that led to economic growth. Distribution was Ricardo's special concern, but not as a dissident from the classical line. Only Marx insisted on the distribution of surplus value and made an insurmountable conflict out of the relationship between capital and labour. It would have been only fair to refer to the latter author or to others who, having entered the Marxist cathedral, consider redistributions or appropriations the way to alleviate social turmoil and unrest. Those with affinities originating in Descartes or Leibniz were first of all interested to understand how the social cake is baked; Marx wanted to know first how it's baked and then why it is unjustifiably unevenly distributed; Piketty is only interested in finding out how it is unequally divided and how inheritance may exacerbate this process.

Secondly, on a skilful selection intended to scientifically inflame the atmosphere. Thus, famous scholars are quoted: Pareto, to exploit the conflict implied by his well-known optimum; Modigliani, from which, following the thread of the pure theory of "life cycle wealth", Piketty borrowed the "useful" idea of the relinquishment of accumulated capital, or of most of it, when one dies; Stiglitz and Krugman constitute exploitable sources for the idea of recourse to the state as an infinite reserve of rationality one may appeal to in order to achieve social justice; Emil Durkheim who, already in 1893, announced that a time would come when ownership rights would cease at death. This context includes a cultural component. Namely, we should know it's not only those who configure Pikettism, via anti-economism, who have something against capitalism, but also those who, in a poisonous manner, inhabit its field. A great deal of famous people is aware of that! For instance, Piketty discretely inserts his great dilemma within the preoccupations of the well-known Balzacian character Voutrin: what standard of living can we achieve through inheritance and what standard of living can we achieve through work? Tolstoy's name proves useful, too! His mercantile and opportunistic Nevzorov is mentioned to exemplify that he "embodies the idea that wealth and merit are totally unrelated: property sometimes begins with theft, and the arbitrary return on capital can easily perpetuate the initial crime" (Piketty, 2014, p. 447). The large framework of a world of intellectuals with remarkable works in such fields as history, ethnology, structuralism, anthropology and sociology, on which the analysis supposedly draws inspiration from, is presented at page 32: Lucien Lebvre, Fernand Brandel, Claude Levi Straus, Pierre Bourdieu, Francois Heritier and Maurice Godelieri. Namely, all these people would participate in the march against capital, inheritance, private property and Piketty's "big-bellied" entrepreneur! Also, John Rawls, Amartya Sen or John Roemer are invoked to justify this type of Pikettist impoverishment of the world, etc.

Thirdly, as an extension of the above, this Pikettist selection also works in a scrupulous manner by avoiding such names, inexorably demanded by the specificity of the analyses, that should definitely have been present. We do not dare say that the lack or the meagre presence of references to Adam Smith, Jean Baptiste Say, Fredric Bastiat, L. Walras, C. Menger, A. Marshall, L. Mises, Hayek, etc., is a sign of ignorance; it is rather part of the manoeuvre Piketty uses to come to a carefully crafted conclusion. This manoeuvre is intended to inoculate the notion that there is no need for microeconomics; nor is there a need for Robert Nozick (1974) to weigh in on the complicated issue of the border between equality of opportunities and that of circumstances as a counterbalance to Rawls' *A theory of Justice* (2010).

Fourthly, the intendedly contorted interpretation and the alteration of meaning of sentences bearing strong institutional force complete the dubious theoretical basis on which the Pikettian construct is built. One example is the first article from the *Declaration of the Rights of the Humans and Citizens from 1789*. From the first proclamation, "People are born and remain free and equal in their rights", Piketty does not learn anything. He is seduced by the accompanying addition: "Social differences may only rely on their public utility". Freedom and equality of rights, not of results, are thrown out to make way for the Pikettian conjecture: "equality is the norm, and in equality is acceptable only if based on "common utility"" (Piketty, 2014, p. 480). There is no link between the starting point and the end point!

Every author is entitled to choose his own sources. There is no objection until it becomes obvious that between what is kept and what is discarded from the references, one the one hand, and the conclusions drawn, on the other, a well-orchestrated game takes place behind the curtains, played according to a tune that originates somewhere else. The issue is that it's not Tolstoy's novels that discuss the problem of the initial theft from which capital accumulation started. Primitive accumulation was the specialty of another "novelist". And, actually, Tolstoy did not place face to face and in opposition something that doesn't fit such logic: capital and merit! Or do we need Vautrin to check whether inheritance somehow transforms us into incurably lazy people? Wasn't Mill closer? Or does Mill also talk about the important role of the elite, a paralysing subject for Piketty?! Amartya Sen dealt with poverty and inequality in a competent manner, as no one else did. But the Nobel laureate gives his solution in *Development as Freedom* (Sen, 1999), considering that liberty is the source of general and personal enrichment. Claiming him as a forerunner while, at the same time, advocating equality by bringing down the peaks towards the bottom is deceitful (Blume & Durlauf, 2015). Claude Levi Strauss was influenced by Durkheim and quoted by Piketty, but he stayed away from the political project and the anti-capitalist atavism of Pikettism.

The failure to draw inspiration from the pioneers in the matters he had addressed, economists with well-established works, exempted Piketty's discourse from dealing with inconvenient questions about his anti-economism, but stripped it of consistency, logic and credible argumentation. Some contact with the *Wealth of nations* and the *Theory of moral sentiments* would have eased the Pikettian fury towards the natural inequalities professed by the liberal economist Charles Dunoyer (1846). Actually, it would have brought additional understanding, necessary to accept the relationship between absolute and relative poverty, and consequently, it would have made some room for the notion that during the entire analysed period, the entire world, in various degrees, fared better and better. Piketty could have learned from Hayek and Mises that social justice has few connections with the morality of results; however, the former is only "honoured" with a footnote, while the latter is totally absent from the landscape. Say, the unexploited Say, would have been very useful. From his elementary but encompassing lesson that "in the offer, the demand is implicit", it would have been deduced that the relationship between these two components of the balancing pendulum is one of mutual causality and that, on this circuit, by means of price, each of the two may be both cause and effect. Namely, it is not only demand that reacts to price, but supply does too. Had this been known, Deirdre Nansen McCloskey (2014) would not have had reasons to impute scientific pains and ignorance to Piketty about his lack of understanding regarding the way in which markets react when a good is in short supply. Something else is just as important. From Say, one could have found out that the entity behind the supply is called an entrepreneur, a character assigned a key role. Piketty talks about the middle class, but does not mention the entrepreneur; however, he doubts the right of risk remuneration. Furthermore, from Say as well as from Smith, one could have found out that the number of people able to start and maintain a large business is not great and that their performance must be rewarded. And on the trajectory of this thought, the bourgeois arrangement would have clarified, not against economism or common sense, but in a natural manner that is hard to accept by all those that have informed Piketty, that the world was created under the sign of inequality, that every opposite approach proved unsuccessful, and the quantitative equalization instituted in the communist paradise proved to be "a pornography", as Christian Michel (1986) puts it—that is, something against nature.

Regardless of what we may call it, bourgeois or proletarian arrangement, it is important to keep in mind that this also means that not every proletarian becomes an entrepreneur. Nobody stops him from becoming one, yet natural inequality prevents him. In addition, the said arrangement also means that two actors shake hands because they are aware that they are not self-sufficient! The division of labour, even a natural one, constrains them to understand their own role or purpose. Here lies one of the main sources of balance. Marx was allergic—he knows why—to the "little vulgar economist Say". Piketty also knows why, and so he obstinately avoided him, proving to be very peculiar when, in 2014, he talked about profit, but did not say a single word about the entrepreneur. However, once one is connected to the sources of the *Capital*, not his but rather Marx's, it is not weird to talk about profit as the exclusive fruit of the capital. The allergy to the entire Austrian School is visible. Mises would have suggested that the difference of appreciation of the future dynamics of business might be an essential source of profit. The modern *Theory of the firm* would also have had something to say about this. Or even James Burnham's (1969) *Era of organizers* would have provided the opportunity to somewhat weaken the Pikettian outbreak against the defying salaries of managers. Solow's (2014) cavalier gesture of comparing the super-managers with the class of rentiers to sweeten the criticism of Piketty deserves appreciation in itself, but does not solve the issue. The problem is to find out whether the economy could work without Burnham's super-organizers who receive "outrageous" salaries for very complex tasks! It would also have been worth quoting from another classicist, namely Smith, that "[i]t is not the actual greatness of national wealth, but its continual increase, which occasions a rise in the wages of labour" (Smith, 1977, p. 103). Namely, what matters is the concern for the growth of the social pie. Afterwards, we shall see how we can divide it. It is only by this growth that we can have our wages raised; the source of such raise does not lie with distribution. How ill-spent these great lessons are! The lack of solid references which raises mistrust in the Pikettian construct may be illustrated with the following sentence from his great opus: "[t]he fact is that all economists— monetarists, Keynesians, and neoclassicals— together with all other observers, regardless of their political stripe, have agreed that central banks ought to act as lenders of last resort" (Piketty, 2014, p. 560). We can only say the following: it is well known that all economists know that not all economists share the same opinion! With unfortunate exceptions.

All economists know that Marx is the "novelist" who tries to demonstrate that the uninterrupted accumulation of capital takes place under the curse of initial theft. Through assumed titles, methodology and cheap tricks he guides us from one level to the next of his 685-page citadel, making all sorts of references except the ones he should have made, but especially by his deceitful way of drawing conclusions, Piketty tries to mimic the "titan". However, he is hindered by a whole bunch of shortcomings in regard to rigor, theoretical foundation, the force of his abstractions and the credibility of his conclusions.

We are interested in Marx because he provides one of the keys for understanding how Piketty wrote his books. Marx is the bulk source of Piketty's own drives that are typical of anti-economism. Thus, the atmosphere of his works is filled with "big-bellied shareholders" and over-satiated people. Pauperization is absolute, since our author wonders: "[w]hat was the good of industrial development, what was the good of all the technological innovations, toil, and population movements if, after half a century of industrial growth, the condition of the masses was still just as miserable as before"? (Piketty, 2014, p. 8). What else is missing from the landscape? Expropriators must be expropriated and chains must be broken, Marx the patriarch would say! And this should necessarily happen because the process of impoverishment is corrosive, says Piketty the apprentice. It originates in the past and goes on into the future! This future is sombre as "[t]he inequality r > g in one sense implies that the past tends to devour the future" (Piketty, 2014, p. 378). And if it devours it, what is there left to lean on? If we give up to the state what we have gathered in a lifetime, and the past works against us in such a manner, what does Mrs. Brundtland think about the relationship between generations? It's hard to say! Not only does Piketty seem to not have heard about the big enrichment of the very period analysed by him, but he takes offence at all reformers inspired by Eduard Bernstein who "(...)had the temerity to argue that wages were increasing" (Piketty, 2014, p. 219). Branco Milanovic, referring to the latest Pikettian work, highlights the concern for the metamorphosis of social-democratic parties from defenders of poor classes to advocates of the rich and educated middle classes and the reorientation of the left towards "La gauche Brahmane" or "La droite marchande" (Milanovic, 2014). Annoyed, Piketty declares himself immunized "against the conventional but lazy rhetoric of anticapitalism" (Piketty, 2014, p. 31), and from this position, he criticizes not only the warmish social-democrats and the workers that embraced them, but also Kuznets. Namely, he thinks that inequalities are always growing and do follow the deceiving shape of a bell.

What else completes the picture of Pikettian anti-economism in this professedly special formula? There are a lot of other places, he believes, that stink of economism! Money has a bad nature; "it tends to grow by themselves" since Bill Gates' assets continued to inflate after he retired! And "[c]apital is never quiet(…) it always tends to transform itself into rents" (Piketty, 2014, pp. 115–116). And we should be very careful with rents because "[d]uring the twentieth century, the word "rent" became an insult and a rather abusive one" (Piketty, 2014, p. 423). Inheritance belongs to the same category; it goes against common sense, allows for a good life without work and cannot be a sign of virtue.

The aberrations around the pretended fundamental law $r > g$ do not actually summarize a discourse on growth. In fact, it is only an opportunity for its author to show himself interested in the dynamics of inequalities and in the way that patrimonies and heritages make them hard to accept. It must also be understood that population growth is not a source of economic growth, but rather a chance "to play an equalizing role because it decreases the importance of inherited wealth: every generation must in some sense construct itself" (Piketty, 2014, p. 83). As old people own most assets and do not intend to die fast enough, young generations have a problem: choosing a profession is "more frequently necessary than in the past" (Piketty, 2014, p. 390). The same conclusions are drawn from the possible decrease in growth. Growth may be slowed down, but what can you do if capital returns increase and this translates to long-term "powerful and destabilizing effects on the structure and dynamics of social inequality" (Piketty, 2014, p. 77)?! Inequalities and assets, they are the problem! Because of them, not even a somewhat natural connection between generations can take form. This is Pikettism!

It is worth expanding the image of Pikettian anti-economism within the normative domain. What solutions are proposed and which model is considered?

Because "such an inegalitarian spiral cannot continue indefinitely" (Piketty, 2014, p. 366) and because "[t]he idea that unrestricted competition(…)is a dangerous illusion" (Piketty, 2014, p. 424) to leave it to the market to solve such a problem, and furthermore, because we have reached a situation in which " (…)the poorest half of the population will generally comprise a large number of people(...)with no wealth at all or perhaps a few thousand euros at most" (Piketty, 2014, p. 258), for all these reasons, all we can do is "shave" the rich and thus calm people down! A tax of around 80% would be welcomed and balanced! Who will assume such a task? Also thinking about possible expropriations, Piketty sees the state as the instrument of salvation. He berates Milton Friedman and his School as they have contributed to a "fostered suspicion of the ever- expanding state" (Piketty, 2014, p. 549). The Friedmanian "limits" must be melted and it must be admitted that "[t]he development of a fiscal and social state is intimately related to the pro cess of state-building as such" (Piketty, 2014, p. 491). A fiscal and social state, pure and simple: this is the economic model worth fighting for! The baffling example at page 572 conveys the notion that the wealth of some people and their productivity are in direct proportion with tax rates, and vice versa! According to estimates, the tax rate for developed countries should be more than 80%! This should not hurt just as "state intervention did no harm"! (Piketty, 2014, p. 99). As the state already exists, one has to only alter its redistributive function and its geographical area! In fact, this model is not national, but global! A globalized progressive taxation system is Piketty's clear and comprehensive norm. If we have fears about the ghost of Pikettism, we may sweeten the pill with the thought that "[t]he progressive tax is thus a relatively liberal method for reducing inequality" (Piketty, 2014, p. 505). This is a strange sample of "liberalism" that cannot be found anywhere else! The pharaohs of Egypt used to send their counsellors to first analyse the flooding level of the Nile and the state of the nation, to ascertain the areas from where earnings could come and the necessary public expenses to be made, and only after that, they set the tax rate: did they comply with the Pikettian paradigm of taxation? Did they have a Mrs. Bettancourt who hadn't worked a single day of her life because she inherited a pyramid, and with Plato's whisper ringing in her ear, did she express her scorn towards wealth, considering it, in itself, a certain source of laziness and corruption? It is hard to believe that anti-economism and adversity to the values that secured the dynamics of the civilized world have such beginnings!

The economic model always possesses a political tegmen. Piketty and all his disciples curse capitalism, but they do not want it totally dead. They want to make it as "rosy" as possible! The professed radicalism is well within the framework of socialism in its French colours. Such frameworks, rather than Marx, are the main sources of the designed model of society with claims of sustainability. The sources indicated at page 32 are only "cultural". With respect to norms, we have to search elsewhere. We are on the territory of Meslier, Morelly or Babeuf. Removing the property, the blind and the ruthless interest that accompanies it, dropping all the prejudices, the horrors, that it supports (Denis, 2016), are impulses that originate in Morelly, not in Levi Strauss. The disappearance of loafers so that the workers can take over the entire stage was the wish of those who utopically preface Piketty, namely Saint-Simon and Charles Fourier. Noteworthy is that Louis-Blanc's "labour oriented society" fits him like a glove. "Big capital as an enemy" was a phrase pertaining to Francois Mitterand (Mitterand apud Peyrefitte, 1983, p. 61).

Socialism, whether in its French or Scandinavian flavour, the latter being the one to which the supporters of degrowth and all ecological economists transparently reference, has not proved perfectly fit for, but also not against sustainability. In the form of the German social market economy, this model also proved resilient and functional, in both social-democrat and conservative nuances. We would believe that this is the model to follow in creating the galactic government charged with progressive taxation that Piketty has in mind. However, we are not led to only consider this line of thought. Piketty leads us into really nebulous and sad areas to find sustainability. These are areas in which Lenin's ideas, which he complimentarily quotes at page 142, have left deep traces. He looks for a society in which the distribution of possession over capital has been "slightly unequal". He is not sure: he just thinks such a thing did exist in the former USSR! "It is difficult to say whether this criterion was met in the Soviet Union and other countries of the former Communist bloc, because the data are not available" (Piketty, 2014, p. 602), he says! Such data could be provided to him by us, the people who did live in the communist bloc. And we can also give him a short answer: that world which lacked data was egalitarian in its entirety. Because it forbade private property almost entirely and directly controlled the distribution of primary incomes, the Soviet Union seems to him desirable, because it was not forced to appeal to income-confiscating taxes. Even though the suggested model was "much less respectful of the law" (Piketty, 2014, p. 602), it has the "merit of consistency" on its side. And, just so we have no doubts about the sustainability of the model proposed by Piketty, here is an epic sample: "the fact that the Soviet Union joined the victorious Allies in World War II enhanced the prestige of the statist economic system the Bolsheviks had put in place. Had not that system allowed the Soviets to lead a notoriously backward country, which in 1917 had only just emerged from serfdom, on a forced march to industrialization? In 1942, Joseph Schumpeter believed that socialism would inevitably triumph over capitalism. In 1970 (...) Paul Samuelson (...) was still predicting that the GDP of the Soviet Union might outstrip that of the United States sometime between 1990 and 2000" (Piketty, 2014, p. 137). What should one think about that? Piketty wrote these lines in 2014. The collapse of the communist bloc had already become history. Placing the Soviet GDP ahead of the American GDP in 2014 verges on the ridiculous. It is also suspicious to endorse the image of a failed state as an inspiring model for the twenty-first century. Schumpeter's prediction was not referring to the totalitarian socialism of the former communist bloc. That is a serious misunderstanding. In 2014, Samuelson, faced with a reality so radically turned upside down, would not have written what he did in 1970. And also: praising Lenin and a system that was very sad and painful for a very large population and giving it an aura is more than hazardous. This risk originates in the unqualified conflation of Western and Eastern socialisms. Socialism is a Western European idea, but the experimentation in all its ugly bareness took place in the East. In the West, it remained consensual and supposedly humanitarian, but in the East, it destroyed countries, confiscated and damaged human values, and filled the "concentration camps" with prisons and Gulags. If the scholars who have no idea how repugnant this experiment was think it cool, civic and even "green" to declare themselves socialist in Europe or in the USA, as fancy as it was to declare oneself utilitarian in the times of Mill, Russell or Keynes, it is a freedom of thought issue. Piketty can be their leader in the march towards the global egalitarian fiscal socialist state. In Eastern Europe, this idea is dead. All arguments are at hand here to nullify the idea of socialism à la Lenin, Stalin, Mao, etc. And this idea is incompatible with sustainability.

In **short**, what should we keep in mind about Pikettism? We underline that:

- Piketty's anti-economism is not new, but is special due to the vivacity and jugglery it uses to be endorsed. Statistical data, "fundamental laws" and shocking interactions between strangely defined values added together in a common soup that mixes, in a hocus-pocus manner, large incomes, inheritances, capital, profit, inequalities, Soviet experiences and the world government are delivered in a manner meant to impress; to illustrate the huge amount of capitalist indifference, the wickedness of accumulation, profit and inequalities compared to the requirements of social justice focused mainly on results and addressed to the inferior centile. The upper centile must only concede. All the rich are ugly to Piketty. Without wondering whether the inheritor of L'Oréal business creates jobs, to him she appears toxic too. The inequality of inequalities of the world defines, in its substance, Pikettian anti-economism.

- The bibliographical selection made by Piketty is fundamentally supported by ideological reasons. The Pikettian scientific work is a small cocoon in a grand ideological shell. The entitlement we are invited to adhere to with respect to the scholarly analysis of thousands of data points pales in front of the utopian project with a recipe of planetary scale that we are also invited to take part in. The socialist good-heartedness with which he builds his plan seems tempting to a certain extent. Beyond that point, caution is advised. We repeat: a model of socialism, in French or Scandinavian flavour, may offer some benchmarks on the path to sustainability. However, when Piketty points towards the former USSR and the Soviet bloc to find a plus of coherence and cadence, then we have a problem not only with the radically proven lack of sustainability of such models, but especially with the sustainability and reliability of such a discourse. For the civilized world that did not experience this type of coherence via the Soviet Union, Piketty's texts may be catalogued under the rubric of the necessary dialog made for the sake of knowledge. However, for the people who have "enjoyed" the "slightly non-egalitarian" atmosphere of the concrete-like coherence of a perfectly and totally organized world, "less respectful for the legal principles", the Pikettian suggestions are an insult. They are an attack not just on sustainability, but also on common decency.
- There is a risk in Piketty's anti-economism being promoted as sustainability under a new doctrine. His great reviewers are the first to lend this trend a helping hand. Solow declares patriarchally that *Thomas Piketty is Right*. He agrees with Piketty in establishing that the economy is able to absorb larger and larger quantities of capital without generating decreasing returns. However, this happens together with the unfair distribution of results. A critical observation of the great economist is that workers can save and invest too. But their gesture lacks a large scope. People who receive

inheritances are favoured in this regard, so Piketty is right again! Not even the global tax on capital seems to disturb him. For Krugman (2014), Piketty is even righter. He finds him unconvincing for his "unintentional and recognized weakness" of explaining the enrichment of 1% of the USA population. He draws attention to the fact that the global tax on wealth is a good idea, but with few chances of success, given the power of the wealthy elite of manoeuvring world politics. As far as the rest goes, Piketty seems to be the perfect scientific child for the recommendation "Capital in the Twenty-First Century is an extremely important book on all fronts. Piketty has transformed our economic discourse; we'll never talk about wealth and inequality the same way we used to". Joseph Stiglitz (2015) is more critical. Using the example of the Rockefeller family, whose wealth should have expanded to $680 billion according to Piketty's logic, compared to $10 billion in reality, the model is disqualified. Criticized are also the relations between the interest rate, the accumulation of capital, rate of return, work productivity, savings, wealth, etc. The conceptual confusion is also indexed.

In short, we mean to say that both the appreciations and the criticisms of Nobel laureate economists relate to the technical aspect, the mechanism and concepts. But no word is said about the doctrinal aberrations of the Pikettian approach. Is there no one disturbed by the so-called virtues of a past system such as the communism of the Soviet bloc? Actually, this is precisely the reason why Piketty's model is not just unsustainable! It is illogical, condemnable and deplorable.

Other reviews we have made reference to are written in the same tone. The concepts, the premises, the mechanism of wealth accumulation and deepening of inequalities, the refusal of marginalistic calculation, the relationship absolute wealth—relative wealth, statistical data, macroeconomics without support in microeconomics, confusion in defining equality, etc.—all very important—are considered. The problem is that Piketty's *Capital* abounds in ideology. He knows it and therefore offered a bookish extension in his work, *Capital and ideology*. Here too, one can find the nodal landmarks of his anti-economism. It is an anti-economism by which the crookedness of capitalism is deduced, in a Marxist style, from the fact that surplus value, after being created exclusively by workers, is the source of profit, rent and other forms of income unpaid to them. But the crookedness is especially outlined due to the fact that the inheritor of wealth feels at ease with morals, even though he does not feel the need to work!

- The generally too inconclusive socialist experiences of the 1950–1970 period stay on Piketty's mind and feed him new ideas and drives. The passion for property, as well as money, could not be ignored by a moralist with his views. He seems distressed that starting with Proudhon, except for communism, which related the religion of material things and property to unproductive consumerism, moral perversion and human degradation, thus attempting to castrate the sentiment of ownership, the other doctrines did not dare go so far. Disappointed by the fact that not even the communist experiment was efficient in diluting the sentiment that success in life cannot be separated from money, goods and properties, Piketty makes a harsh indictment of "propertarianism", a phenomenon that is nowadays manifest in the former communist countries. He does not comprehend how in these countries, forced to the most extreme level of wealth equalization, an appetite for goods, money and properties to compensate for a confiscating past is still possible! Yes, we can say it is really possible and occurs rapidly, beyond any moral restriction and in accord with human nature too long crushed by an aggressive regime. In the preface to his book, Stephen Young (2003) recommended that "it would be welcomed that today's Romania comes back to its values and traditions extant before the communist regime. All the things that had once given people force and confidence may become valid again today, conferring them the necessary guidelines in making right decisions regarding free market transactions". To make right decisions regarding free market transactions, one needs to be the owner. Otherwise, one can only trade with stolen goods. Mr. Young understands this perfectly; Mr. Piketty doesn't as long as he is disturbed by "propertarianism". There is something else to say in this regard. For the opposers of the communist regime in the countries referred to by M. Young, the expropriation by forced nationalization and cooperativization was accompanied by the "total confiscation of wealth". This meant that the regime confiscated the house and all goods from the family of an enemy of communism. We invite Mr. Piketty to hold a discourse on propertarianism to the inheritors of such a family! A sustainable discourse!

- The culture of property and money does not include anything hypocritical. Blaming and extinguishing this culture is equivalent to an undermining of the fundamental components of success. The courtyard of the moneyless and property deprived poor is not the most adequate place to define success in life. If there is anything left to talk about, it relates to the manner in which goods have been gained and used. In addition, property is an institution with a strong historical meaning. Combatants in wars were rewarded with properties as a thank-you gesture from the nation, but also as a form of deserved justice. War was the opportunity. The same was intended by the secularization of monastery wealth. On the same line of historical investigation, it was ascertained that the issue of property is essentially a problem of productivity. And it was also ascertained, without clear counterexamples, that ownership with an individualized master—private property—is infinitely better managed and generates much higher returns than public property. Consequently, within the distinction between a "lazy" owner by inheritance and an "honest" owner by work and personal accumulation, the criterion of returns should split the difference. If we have all admiration for an individual who, as entrepreneur, produces goods and employs workforce, and we are not very much disturbed by the fact that, in doing so, he becomes rich, why should we get so mad and anathemize his inheritor who continues the business directly or by delegating the management to move the inheritance forward by conserving and enlarging it, maintaining or even increasing the number of jobs? Laziness should only be discussed by moralists, just within the boundaries of the Law, if the inheritor squanders the inheritance, thus stimulating for a while a consumption to whose sustainability he has not contributed. How should a law in this matter look like, a different law than Piketty's "Law against snobbism"? Mill suggested it. Marx insisted. Piketty does too, but reality shows that it is hardly possible to re-establish every time equality at the starting point. A socialist-communist cliché—a current photo—may be confusing. "Today" we ascertain unjustified inequalities in wealth and we redistribute till we bring everyone on the same level. Some people's pluses go towards benefitting other people. And then? It is the dynamics—the film—rather than the photo that reminds us that inequalities occur every day. We may only suppress this tendency in a totalitarian communist regime that levels human minds and behaviours to the point of annihilation. Actually, before starting such a highly utopian project as the one proposed by Mill, Modigliani, Durkheim or Piketty, we may listen to the voice of a liberal, a French liberal economist, Henri Lepage, who says in 1989 that "Property does not mean "the right to do what you want with what you have" but the right to freely decide to use resources provided that this does not infringe on the similar rights of others" (Lepage, 1989). Free market does exactly that: it confers the right to decide to be efficient, because otherwise, what you have will be transferred to the person possessing a more skilful mind or pair of arms. There is no need for Lenin, nor the coherent model of the former Soviet Union. That was not the place for sustainability and resilience. And a liberal economist, as we have seen, tells us that we no longer need to stumble against property as a natural and inalienable right. Actually, property should be put to work. If it isn't, one becomes useless and joins the "reserve army" of Piketty's loafers!

- With a fire extended on thousands of pages, Pikettism hopes to go a long way. It is hard to believe that it will become a doctrine. It needs a solid and original theoretical foundation to stand on, and while it claims to have it, it doesn't. We believe it will only be a candidate to the status of a "destructionist" current in Misesian sense (Mises, 1981), set against the traditional definitory values of Western capitalism: liberalism, private property, personal wealth, profit, inheritance, free market, individualism, etc. A product of the original country of socialism, Piketty does not claim other mentors, such as Gramsci, Lukacs or Marcuse. He rather operates with "big-bellied" characters and workers, remaining under Marx's wing in trying to find the "objectivity"—operating through fundamental laws—of the conflict between labour and capital. Cultural Marxism via the School of Frankfurt seems too gentle to him. At the same time, he does not seem to be contaminated by the green watermelons group (DiLorenzo, 2016) gathered around Latouche, itself infused with strongly socialist references and projections. Namely, Piketty does not invoke, as Heilbroner (1990) does, the need to return to socialism as the only way to solve our environmental problems. In contrast to the American economist sporting socialist views, Piketty does not seem willing to admit in any way the failure of the communist economic organization, and once the mistake is admitted, to plead for a new trial. Socialist experiments "not sufficiently conclusive" do not suggest failure, but rather weakness. His pursuit includes a plus of radicalism in criticizing capitalism, wherever it may be, in order to sweeten it and fill it with red roses. This line of thought, benefiting from scholarly support, is indisputable proof that Piketty is a great destructivist. His concern to fix the world by using the French socialist mantra, a system championing the number of taxes and duties, bolstered by a world government that should confiscate wealth excesses—this is the blood-red ghost of Pikettian anti-economism! We believe that the painful denial of Pikettism has not been, and it will not be, essentially academic. Enjoying high-level academic support, Piketty only fills his writings with exemplary phrases. The reality of a world who "conclusively" and "coherently" experimented with this system will tell him how much utopia and lack of sustainability can be found in the world republic and the equalizing, deeply humanist aims of the world pictured in his *Capital in the XXI century*. In forbidding birthright and personal wealth, Piketty hopes to foster smiles. Redistributions, confiscations and anti-propertarianism might cause some ovations. However, in the long run all of these disappear. Resilience refuses them.

4.5 Concluding Remarks

The title of this chapter is meant to announce, in itself, our intention: if we do not pay attention and we make social concerns the alpha and omega of theoretical discourse and practical approach, our lesson in economics might amount to nothing. This warning does not come without reason, on the contrary. As a means and an aim of offering a proper place to produce loud "scientific clamor", social concerns are only competing against the environmental footprint and its adjacent frights. In short, these are our conclusive opinions on the subject:

- The economy has always been connected to politics. Which one is more important? This is the question that rummaged for and offered answers impregnated by the ideology that gave them birth. On its merits, this relation should not have anything to do with ideology. However, the economy is highly important for the life of the city, but it does not take decisions. Politics does. And because it decides, it lends its colour to economics. It does so forcefully, under autocratic regimes, and voluntarily under democratic rule. In the latter cases, it is important for democracy to be assisted in valuing its selective characteristics: it should see that its most valuable members are engaged in macro-management activities. Otherwise, it fails and provides social and environmental activists with arguments for a plus of authoritarianism in solving the problems of their area of interest. This is the source and the core of a perversion. According to the Smithian school, the state is democratic, minimal, but powerful; it does not leave important decisions for others to make. The state of social welfare, considered more important than economic welfare, is no longer a powerful state. It is seen as a reservoir of wisdom and a final saviour for those who it allows to survive when they should be left to go bankrupt, while assisting the deaths of those who deserve to live. Its main colleague in this mission is the central bank. Living with it as concubines, they corrupt each other's roles, and afterwards, thus equipped, they corrupt the economy.

- Economics has been weighed against social concerns, and for some time now, also against the environment. The axiom that links the three elements of sustainable development seems to be the following: without social and ecological peace, there is no economic peace. This means that economics must take care of social and environmental issues, even before taking care of its own business. We have tried to show that every hierarchy in this matter is invalid. If there is any order of importance of the said domains, it cannot be established by the economist, the ecologist or the social activist. Theory can only ascertain that it is the economy that feeds, shapes or ends the other two domains. At the same time, good theory has argued that the economy in itself is a logical inadequacy. And the founders we have been talking about have produced good theory. No classicist has attempted to put social concerns before the economy and then try to answer absurd questions. The fact that the economy must serve society was an axiom of their analyses. If there are back-and-forth issues within the social embroidery they used to stuff their work with, they concern the relationship between individual good and social good. They did not try to melt individual welfare into general social welfare. The exceptions are Mill and especially Marx. However, it is worth remarking that their first concern was the economy. It has to work well before there is any object of dispute on social, distributive and redistributive levels. The neoclassicists did not proceed differently. They said that economic efficiency is nothing unless it also contributes to the betterment of society. They said it, admittedly, as a late consideration, after having filled page after page with heroic scientific abstractions. But in the end, when they tackled social issues, and some of them, not few in numbers, environmental ones as well, they proved that they hadn't neglected the other domains, quite the contrary. We mean to say that every criticism of the founding classicists or neoclassicists under the charge of a lack of interest in social or environmental issues is just gratuitous!

- The perspective of the dialectical relation between the economy and society, up to the point when economic peace no longer counts for much in defining sustainability, has followed a sinuous route and has included multiform factors, sequences and currents of ideas.

First of all, doctrinal transgressions committed under the influence of novel facts against the status quo have raised the quota of social issues within the range of factors that contribute to macro-decisions. The Keynes moment is one of them. Through it, social issues gained a level of priority that is perverting on multiple levels. We are talking about positioning the state in an unnatural place in the causal chain that links production to consumption; about the artificial gains of primacy that distributive justice scores against the efficiency of production; about the birth of the dubious logic of creating jobs before any production takes place, etc. Plentiful examples of this sort are to be found in the text. We add here that the moment also launched a fashion of out-of-context quoting an economist, often to take advantage of that scholar's notoriety, with the purpose of emphasizing the prevalence of social issues at any price; when the street "catches on fire", we should leave economic laws aside. Authorized voices have said it. There is no place to turn back.

The claim to transgressions, if not with respect to doctrine, at least related to object and method, also comes from those who, dissatisfied that standard science has no ad hoc answers to environmental issues, plead for another science. Ecological economics brings to the table a mixture of social and ecological issues—so that, after criticizing economism, it may turn our face towards nature. Never have the advocates of social concerns found better suited scientific buttresses to climb through the ranks and call the shots! If carefully analysed, the physiocracy of Ecological economics is nothing but a phony physiocracy. It is obvious that the philosophy of natural order stands in its way. When it devises its projects, it finds too much democracy to be inadequate. It needs resources and funds for non-polluting technologies and the cleaning of the planet. The eco anguish administered by the left under the waving flag of social issues needs money to solve its problems, arguing from the position of an economic science fuelled by Wikipedia. And it is from this position that it is believed the economy should take a break. During this break, we can quietly take care of redistribution and social justice.

- It was not only severe gaps in the realm of facts that led to the metamorphosis of the state into an applauded redistribution machine and the insemination of an attitude of manic revendication. Institutional initiatives meant to democratically revolutionize social dynamics contributed as well. We have outlined in the text the "collateral" consequences of the Declaration of Independence and the Declaration of Human Rights. Both, but especially the latter, impregnated the world in an institutionalized and governmentally approved manner with the morbus of absolute utopian equality. Such "edicts", and they are not the only ones, form the authorized sources that fuel the social gas tank, ready to explode whenever a social group reevaluates the utopian form of egalitarianism. An invitation from a scholar with authority, possibly awarded the Nobel Prize, for the workers to democratically participate in the decision-making process of a company stokes once more the flames of social pressure.
- In the fighting unit conceived to degrade economics in its relation to society, a special place is occupied by the artillery pointed against economism. Economism is perceived as a philosophy with a mercantilist flavour which eulogises profits, industrialism, productivism and material wealth. All these are associated with the neoclassical model of the free market, accused of neglecting inequalities and serving as the slave of a hideous and unnatural hedonism.

  The campaign against economism, that goes hand in hand with the post-war model of capitalism, stands chances to deliver a devastatingly corrosive blow. This movement is supported by noteworthy names of economic science, but also by institutions at the global level. The Brundtland Report was not conceived in this sense, but it was enough that its wording specifies that "profit is not everything" for the said document to be invoked as a powerful argument backing all the enemies of profit. Anyway, the affirmation that comes across from this world of ideas is that it's à la mode to be against economism. It can even be attached a hashtag: #IresistProfitOrIDieFromPureHappiness!

We do not deny that in all this mess in which economic laws are misconstrued and turned upside down to be negated, something positive goes on as well. We believe anti-economism may be saluted when, thanks to it, the statistics of macroeconomic indicators are subjected to severe and justified criticism. We refer to the inculcation of the illusion of welfare growth created by means of deceitful GDP statistics. In other words, we salute the invitation to submit statistics to an exercise of sincerity. It is an invitation to find out using its supply of data whether GDP pluses are also conducive to human development and welfare for all or they just benefit a tiny fraction of the population; whether a country grows by consumption or rather by production; whether the main source of social security expenses should come from taxes and duties; whether poverty may increase alongside growth or whether foreign investments represent a universal panacea or not, etc. To ask for socially determined statistics, not only of the economic kind, can only be thought of as a good sign. However, such a plus attributed to anti-economism is seriously weighed against the unsound persecution against profit and the "successful" perversion, academically packed, achieved in the domain of redistribution and social justice.

It sufficed that Milton Friedman, disarming up to the point of defiance, claimed that the sole purpose of the company is profit, for a real cannonade to be let loose. It is well known that the great economist was no pioneer in supporting this opinion. Probably, his exclusivism in establishing the nature and the purpose of the company bothered more than his phrasing of a truth known to everyone, but only conceded by some. Could he have gathered several adherents had he claimed that the company has, first of all, social reasons, and that it is meant to serve public interest? Probably so, but then Friedman could not have spoken from the position of leader of the Chicago School, but from the position of some nameless school in some communist country. Friedman and his predecessors, the classical and neoclassical founders, never thought

> about profit in and of itself, bereft of the social finality of production, whose result it actually is. But it is clear that they thought as one that without profit there is no entrepreneur and, thus, there is no capitalist company. Period! Whoever goes on beyond this point sells a tongue-in-cheek discourse, deceiving a world that has no idea that Marxist socialism irremediably failed, mainly because it removed profit from political and economic philosophy, practically destroying the institution of entrepreneurship.
>
> The fight against people who allegedly see nothing further than mere profit only serves as a banner for a larger war. Its front is opened, facing towards such feared "enemies" as income inequalities, individualism, "thick-skinned" inheritances and social injustice. Reports signed by notorious names, see the Stiglitz-Sen-Fitoussi Report, try and unfortunately succeed in disarticulating fundamental economic judgements, thus putting the cart before the horse. Consequently, we find out that distribution is more important than production, and that we should first deal with incomes and consumption, health and quality of the environment and only afterwards should we turn our attention to industries. Making GDP the main focus of our attention is as disturbing as its unequal distribution in the world. On the whole, we find out that we do not need growth to get to heaven! By having a distribution subordinated to the levelling will of the masses, we can solve all the questions concerning health, poverty, education and happiness!

As an able "child" of this world of ideas, Thomas Piketty proves highly efficient in these matters. This is why we have assigned him a special place in the contents of this book. In these conclusions, we insist that references to him are meant as a synthesis of this entire movement of ideas that gain a wide scope within the domain announced by the very title of this chapter: the pressure of the social component, driven by a disagreement with preponderantly ideological and populist stakes, risks to invalidate, maybe unintendedly, a founding lesson about the sustainability of an economy.

References

Atkinson, A. B., Piketty, T., & Saez, E. (2011). Top incomes in the long run of history. *Journal of Economic Literature, 49*(1), 3–71.

Benabou, R. (1996). Inequality and growth. *NBER Macroeconomics Annual, 11*, 11–74.

Birchall, I. H. (1997). *In The spectre of Babeuf*. Macmillan Press.

Blaug, M. (1985). *Economic theory in retrospect*. Cambridge University Press.

Blume, L. E., & Durlauf, S. N. (2015). Capital in the twenty-first century: A review essay. *Journal of Political Economy, 123*(4), 749–777.

Brecher, R. A., & Gross, T. (2017). Unemployment and income-distribution effects of economic growth: A minimum-wage analysis with optimal saving. *International Journal of Economic Theory*.

Burnham, J., Claireau, H., & Blum, L. (1969). *L'ère des organisateurs*. Calmann-Lévy.

Dasgupta, P. (2008). Nature in economics. *Environmental and Resource Economics, 39*(1), 1–7.

de Sismondi, J. C. L. (1838). *Études Sur L'Économie Politique: Tome Second*. Treuttel et Würtz.

Denis, H. (2016). *Histoire de la pensée économique*. PUF.

DiLorenzo, T. (2016). *The problem with socialism*. Regnery Publishing.

Duflo, E., Kremer, M., & Robinson, J. (2008). How high are rates of return to fertilizer? Evidence from field experiments in Kenya. *American Economic Review, 98*(2), 482–488.

Duflo, E., Kremer, M. & Robinson, J. (2011). Nudging farmers to use fertilizer: Theory and experimental evidence from Kenya. *American Economic Review*, 2350–2390.

Dunoyer, C. B. (1846). *De la liberté du travail ou Simple exposé des conditions dans lesquelles les forces humaines s'exercent avec le plus de puissance*. Liège: Libr. Scientifique et Industrielle, de A. Lerouz.

Facchini, F., & Couvreur, S. (2015). Inequality: The original economic sin of capitalism? An evaluation of Thomas Piketty's "Capital in the twenty-first century." *European Journal of Political Economy, 39*, 281–287.

Foucault, M. (2005). *The order of things: An archaeology of the human sciences*. Routledge.

Friedman, M. (1962). *Capitalism and freedom*. University of Chicago Press.

Friedman, M. (2007). The social responsibility of business is to increase its profits. In W.C. Zimmerli, M. Holzinger & K. Richter (Eds.), *Corporate ethics and corporate governance*. Springer.

Friedman, M., & Friedman, R. (1980). *Free to choose: A personal statement*. Harcourt Brace Jovanovich.

Galbraith, J. K. (1975). *Economics and the public purpose*. New American Library.

Georgescu-Roegen, N. (1971). *The entropy law and the economic process*. Harvard University Press.

Gorz, A. (2008). Richesse sans valeur, valeur sans richesse. *Cadernos IHU Ideias, 31*.

Harribey, J. M. (2010). Elements for a political economy of sustainability based on a decline of goods. *Revue Française De Socio-Économie, 2*, 31–46.
Hayek, F. A. (1960). *The constitution of liberty*. The University of Chicago Press.
Hayek, F. A. (1976). *The mirage of social justice*. The University of Chicago Press.
Hayek, F. A. (1982). *Law, legislation and liberty: A new statement of the liberal principles of justice and political economy*. Vol. 2 The mirage of social justice. Routledge & Kegan Paul Ltd.
Heilbroner, R. (1990). Reflections: After communism. *The New Yorker*, *10*.
Jackson, T., & Victor, P. A. (2016). Does slow growth lead to rising inequality? Some theoretical reflections and numerical simulations. *Ecological Economics, 121*, 206–219.
Jouvenel, B. D. (1972). The language of time. In*The future is tomorrow*. Springer, Dordrecht. https://doi.org/10.1007/978-94-010-2826-4_19
Keynes, J. M. (2019). *The economic consequences of the peace*. Palgrave Macmillan.
Keynes, J. M. (1963). *Essays in persuasion*. W.W. Northon & Company.
Keynes, J. M. (2018). *The general theory of employment, interest, and money*. Palgrave Macmillan.
Krugman, P. (2007). *The conscience of a liberal*. W.W. Norton & Co.
Krugman, P. (2014). Why we're in a new gilded age. *The New York Review of Books*, 8 May. https://www.nybooks.com/articles/2014/05/08/thomas-piketty-new-gilded-age/
Kwak, J. (2017). *Economism: Bad economics and the rise of inequality*. Pantheon Books.
Lenin, V. I. (1900). *"Left-wing" communism: An infantile disorder*. Progress Publishers.
Lenin, V. I. (1951). *Economics and politics in the era of the dictatorship of the proletariat*. Foreign Languages Publishing House.
Lenin, V. I. (1961). A talk with defenders of economism. *Collected Works, 5*, 313–320.
Lepage, H. (1989). *La "nouvelle économie" industrielle*. Hachette.
Marinescu, C., Glăvan, B., Staicu, G., Pană, M. C., & Jora, O. D. (2012). *Capitalism. The logic of freedom*. Bucharest: Humanitas
Marshall, A. (2013). *Principles of economics*. Palgrave Macmillan.
McCloskey, D. N. (2014). Measured, unmeasured, mismeasured, and unjustified pessimism: A review essay of Thomas Piketty's capital in the twenty-first century. *Erasmus Journal for Philosophy and Economics, 7*(2), 73–115.
McGee, R. W. (2009). Some overlooked ethical aspects of bailing out banks and the philosophy of Frédéric Bastiat. In *The banking crisis handbook* (pp. 597–612). CRC Press.
Michel, C. (1986). *La liberté: Deux ou trois choses que je sais d'elle*. Institut économique de Paris.

Milanovic, B. (2014). The return of "patrimonial capitalism": A review of Thomas Piketty's "Capital in the twenty-first century." *Journal of Economic Literature, 52*(2), 519–534.

Mises, L. (1981). *Socialism*. Ludwig von Mises Institute.

Nock, A. J. (1943). *Memoirs of a superfluous man*. Harper & Brothers.

Nozick, R. (1974). *Anarchy, state, and Utopia*. Basic Books.

Peyrefitte, A. (1983). *Quand la rose se fanera*. Plon.

Persson, T., & Tabellini, G. (1991). *Is inequality harmful for growth? Theory and evidence* (No. w3599). National Bureau of Economic Research.

Persson, T., & Tabellini, G. (1992). Growth, distribution and politics. *European Economic Review, 36*(2–3), 593–602.

Pigou, A. C. (1920). *The economics of welfare*. Macmillan.

Piketty, T. (2014). *Capital in the twenty-first century*. The Belknap Press of Harvard University Press.

Piketty, T., & Saez, E. (2003). Income inequality in the United States, 1913–1998. *The Quarterly Journal of Economics, 118*(1), 1–41.

Piketty, T., Saez, E., & Zucman, G. (2018). Distributional national accounts: Methods and estimates for the United States. *The Quarterly Journal of Economics, 133*(2), 553–609.

Piketty, T. (2020). *Capital and ideology*. Harvard University Press.

Pullen, J. (2016). Malthus on growth, glut, and redistribution. *History of Economics Review, 65*(1), 27–48.

Rawls, J. (2010). *A theory of justice*. Universal Law Publishing Co Ltd.

Roberts, R. (2015). *How Adam Smith can change your life: An unexpected guide to human nature and happiness*. Portfolio.

Samuelson, P. (1955). *Economics* (3rd ed.). McGraw-Hill.

Sen, A. (1999). *Development as freedom*. Oxford University Press.

Smith, A. (1977). *An inquiry into the nature and causes of the wealth of nations*. University of Chicago Press.

Smith, A. (2004). *The theory of moral sentiments*. Cambridge University Press.

Solow, R. (2014). Thomas Piketty is right. Everything you need to know about capital in the twenty-first century. *New Republic, 22*.

Stiglitz, J. E. (2012). *The price of inequality: How today's divided society endangers our future*. WW Norton & Company.

Stiglitz, J. E. (2015). The origins of inequality, and policies to contain it. *National Tax Journal, 68*(2), 425–448.

Stiglitz, J. E., Sen, A., & Fitoussi, J.-P. (2009). *Report by the commission on the measurement of economic performance and social progress*. https://ec.europa.eu/eurostat/documents/8131721/8131772/Stiglitz-Sen-Fitoussi-Commission-report.pdf

Tollison, R. D., & Buchanan, J. M. (Eds.). (1972). *Theory of public choice: Political applications of economics*. University of Michigan Press.

UN General Assembly. (2000). *United Nations Millennium Declaration, Resolution Adopted by the General Assembly* A/RES/55/2. https://www.refworld.org/docid/3b00f4ea3.html. Accessed 30 June 2021.

Walras, L. (1954). *Elements of pure economics, or, the theory of social wealth*. Thomas Nelson & Sons.

Walras, L. (2010). *Studies in social economics*. Routledge.

Young, S. (2003). *Moral capitalism. Reconciling private interest with the public good*. Berrett-Koehler Publishers.

CHAPTER 5

Degrowth—A Logical Inadequacy?

Many years have passed since the first signals that opaque clouds might question the issue of increasing production and wealth. 173 years since J. S. Mill wrote the famous fragment on stationarity! 156 years since Jevons articulated concerns that coal might run out! 49 years since the *Limits of growth* and 34 years since the Brundtland Report that made the concept of sustainable development prominent! And so what? No government has ever proposed to bring the economic engines to a standstill. Oil hasn't run out either. Instead, durable development has intensely lived its moment. It stirred spirits, opened new leads, caused thousands of debates, called for the creation of new sciences and changes of paradigms; it produced tons of literature and raised the Hirsh index to levels that are suspicious to many economists. At the same time, growth seems to stay in favour and even come out strengthened after its confrontation with a pathological ideology that has a strong emotional component—degrowth. It is difficult to find a louder episode in the dynamic of economic science than the degrowth moment. It was a moment that occurred and was fed by the inherent contradiction of a science made to solve the ingrate equation of resources and needs. This moment exaggerated limits and diminished chances, it translated into a troubling oxymoron that absorbed many great intellects, suggesting the need to find convincing arguments that development might only be sustainable by means of degrowth!

I. Pohoaţă et al., *The Sustainable Development Theory: A Critical Approach, Volume 2*, Palgrave Studies in Sustainability, Environment and Macroeconomics, https://doi.org/10.1007/978-3-030-61322-8_5

Can we be won over by this notion? Yes and no. YES, if we enjoy scientific revolutions for the sake of revolutions; if we are sensitive to shocking words and decreed propositions; if instead of adding to the already existent hardcore, we propose dismantling procedures; also, if by changing our language and thinking mode, we blow our own horn, claiming that the great winner of economic dynamic may be the economy that decreased the most! NO, if we do not forget that through a philosophy that has become classical, most of the world has eradicated starvation; if we do not take the risk of converting our society into a socialist-communist one by renouncing a paradigm that, with all its shortcomings, has helped create wealth.

5.1 The Source of the Confusion: Consolidating Scientific Props!

The principled answer to the affirmation in the title is that the primary warnings are to be found in the classical "stationary state". They are not so much to be found in the pessimistic propositions of Smith, Ricardo and Malthus as in the few pages from Mill's *Principles*. We have tried in the 1st volume to convince that not even the rest period after sustained labour can find a logical place in the schemes of governmental policy. At the level of the individual, it can! Other than that, students of economics learn about the stationary state under the label of entertainment during lessons on equilibrium or as a working hypothesis for a model. This is all, because both the state of equilibrium and stationarity only exist as theoretical constructs. Quoted ad nauseam in counterbalance to the hundreds of pages that speak for the brilliant classicist that was Mill, a creator of good theory, the lines on the stationary state have fuelled and offered support to most attempts at scientifically validating degrowth, however, sane or insane they may have been. They were good and interesting when, fructifying Mill's notions, the supporters of degrowth or zero growth look for the spot in which material progress should make place for moral and intellectual progress, provided anything like that is possible. It's just that Mill did not go that far. In the long run, his idea is translated into economic dynamics. He gives the impression that true progress, namely moral and intellectual one, only begins after material progress has reached a phase of saturation. He did not talk about the instant when one should settle down to allow for moral progress to assimilate what is to come. Mill only launches the idea and then falls into deep daydreaming. And this happens only to suggest that while fantasising or even sleeping, one does not have to eat, and after getting up one feels distanced from everything and only finds out from other people that he is in a post-industrial society "based on human development rather than material acquisition" (Raskin et al., 2002, p.18). We do not wonder that Mill's socialist concoction pushes him towards Rousseau-like contemplations, where in a warm and welcoming atmosphere with good and wise people around, bound into brotherhood through unseen threads, he enjoys introspections while tasting the joy of universal idleness. However, we do marvel at the painstaking rush to differentiate between what amounts to mystery in Mill's work and what is imitable that may be used in such a risky attempt as degrowth. We are also surprised at the possibility of seeing in Mill's project "a knotty tangle of determinist and utopian argumentation" (Dale, 2013, p. 438). Mill's determinism, according to Dale, would amount to a "benign" drop of the profit rate up to the point of its stationary standstill! What kind of invisible hand would do something like that, Smith cannot say! Also, nobody knows what would make the entrepreneur accept such concessions! However, in order to support the scientific madness of a need for degrowth, such a notion may be appealing.

The possible gaps caused by the limitations identified by Smith, Ricardo, Malthus and Jevons have not remained dormant either. They were added to, in a catastrophic manner, by N. G. Röegen. He is neither the first nor the last to state that our planet is limited. Yet, he does so using the means of the scholar, linking this issue to thermodynamics and the horrid entropy, to implant into our brains the unforgiving notion that a piece of burnt coal is transformed into ash, that there is no turning back, just as an omelette cannot be turned back into eggs, not even by using some sort of arithmomorphism! His writings, in offering support to the excessive credibility of any sinister prophecies (Georgescu-Roegen, 1971, 1976), would lay the fundamentals for the new Scriptures in this field. He would be followed by all ecological economists, non-economist ecologists, environmentalists, environmental activists as well as all reasonable and purebred prophets of doom, who would stand in line to adopt his ideas and would lose their sleep in the process. Summoning new sciences to befit their ideas, they would forget about the classicists, castigate the neoclassicists and embrace Marx and Engels when they should and, especially, when they should not. They, especially do so when they should not because it is difficult to catch any of them in a "stationary position". The permanent advance of production forces with respect to production relations, the Marxist dialectical toy, pertains to movement. The Promethean human dominance, the fact that humans are supposed to master nature, permeates the logic of the discourse of both of them. In fact, the warning Reports of the Club of Rome cannot be accused of Marxism. Marx does not enter there; however, Malthus does get in, and it is also him that comes out. The reasonableness of the Brundtland Report is given precisely by the intended doctrinal equidistance behind it. It does not invoke any ideology and is intended as an objective analysis of the status-quo, of the limits and opportunities for sustainable growth.

The fear of resource finitude has brought a lot of ghosts to haunt economic science. And the shocking solution that everything might get solved through degrowth coagulated the attention just like the miracle from Cana, when Jesus transformed water into wine. Especially, if tutelary authorities or voices from the tribunes decree that everything would be all right, and by chasing away crises, unemployment, and inflation, we would find again our composure and traditional values; that by "getting out of capitalism and the world of material stuff", we can provide everyone with three meals, thrice a day; that by eliminating the fourth meal from the table of the enriched overweight, by inscribing the right to laziness in the constitution, by satisfying Lafargue and befriending the woods in sweet poetry, we may gain votes! And where the authority of ideas fails to come into prominence, well-crafted and sophisticated words may fill the void, and a magic wand in the hands of a man skilled in creating a new vocabulary may realign things. In fact, it would align in bulk, as it is eclectics, not linearity, that brings life into movement. It is a frisk movement, considering that degrowth is not a reaction to standard growth, but especially to durable growth, a three-dimensional concept concerning economic, social and environmental aspects. However, degrowth revolts on all these three levels; it doesn't let anything go.

The largest front opens in relationship to the environment. The categorization of the movement animators into "three positions", according to their stance on environmental issues made by Olivia Montel (2017) is well suited when it comes to deciphering the level of dispersal with which it is believed that the anti-Economics front may be served. Firstly, standard environmental economics; then, ecological economics; finally, degrowth. The first is considered a science of neoclassical origin; the second—a whole new science; the third—a pamphlet! We contribute nothing new here with respect to the criteria that distinguish them. We only think it is interesting that ecological economics accepts sustainable development, while the supporters of degrowth loathe it!

We shall focus our attention on the last two items of this classification, as they are interconnected. Precisely for this reason, they are difficult to differentiate and catalogue in a clear manner. Herman Daly and Josh Farley seem to use Roegen, especially with respect to the rigour and the nature of the analysis apparatus. As consecrated points of support for the young science of ecological economics, David Acemoglu and James A. Robinson are, especially brought up to highlight the institutionalist dimension of a possible paradigm. Röegen, Perroux and Schumpeter are evoked by Constanza in his effort to provide ecological economics with foundations. Amartya Sen seems interesting to them with respect to and by means of his "capabilities" in defining the new paradigm and in calculating a new GDP. Kuznets is not forgotten by any GDP critics.

In opposition to ecological economists, Latouche, René Passet, and the entire constellation of the French from *La Décroissance* (Cheynet, Schneider, Aryés, etc.) present themselves as another ramification of Roegen's discourse, but they, especially make use of the normative part of his work. In the same perspective, they reclaim Mill as forerunner to show that it is proper to stop and to be thankful for what we have; Pigou with respect to the feeble effect of externalities; Max Weber to ascribe to him the harmfulness of the "spirit of capitalism"; Galbraith with his "satisfied majorities"; Veblen to invoke the perverted substrate of the imperialism of the pro-acquisitive social code; Braudel to theorize the harmful capitalist alienation; Jevons to claim that it is time to suspend material wealth, also him for the "rebound effect" noted by all the scholars interested in substantiating a paradigm; Stiglitz for the relevance of GDP criticism; or even Hannah Arendt to find out that even a great philosopher thinks about the necessary time for contemplation! (Latouche, 2009).

The social component is not exempt from the degrowth revolution, either. Capitalism, with all its companions—free market, profit, and ostentatious consumption—make the object of critical analyses from positions that are allergic to the Austrian school, the Chicago school or any other economic school that makes accumulation and profit the axis of their economic dynamic. Ivan Illich, Jacques Ellul, Andre Gorz and Cornelius Castoriadis are just a few of those that produced works on the subject. Ignacy Sachs also has problems regarding capitalism and does not view it as compatible with "eco-development". He even invokes Gandhi to motivate his break with consumerism to reach a more rational society.

The struggle against economism cannot avoid the country of Lenin, the first user of this concept, and all his contemporary critics of the industrial revolution. Arghiri Emmanuel and Samir Amin also consider that the unequal exchange has no cure! Etc.

It is hardly possible to make a canonical and exhaustive review of the degrowth literature. The number of scholars is overwhelming. The above are just defining excerpts for the framework within which this subject is being discussed. It is a framework in which eclecticism and multidisciplinary approaches are definitory. A variety of the sources and, especially authoritative names are brought to the analysis as crutches and patches. The supporters of degrowth claim that their field also includes "holy fathers" they may count on and whom they follow in their works. They also want to advance their work. It remains to see whether the *Farewell to Growth* of Serge Latouche or the *Law of entropy* of Georgescu-Röegen would really have the force of Scriptures in transmitting such messages as these authors claim to have the power to change the world. To go beyond material and step into imagination, taking economic science with them!

5.2 New Meanings in a New Language

The question "What is degrowth?" is answered both by those who impose the concept and those who refute it. With the latter, one may even detect a plus in clarity due to a plus in ease.

When one does not know something, one appeals to a person competent in the matter at hand. The scholars who attempt to clarify the term are not few. However, the "leader" knows best. Better, more straightforward and more interested in giving a meaning not exposed to criticisms is Serge Latouche. Between pages 7 and 9 of his work, *Farewell to Growth*, he tries, by leaning on Paul Aries and André Gorz, to elucidate the concept intended to be turned into the hardcore of a new paradigm. The entire text entitled *What is degrowth?* is worthy of quoting. A short excerpt should include the following main sentences: (1) "'De-growth' is a political slogan with theoretical implications, or what Paul Ariès (2005) calls an 'explosive word' that is designed to silence the chatter of those who are addicted to productivism" (Latouche, 2009, p. 7); (2) "De-growth is not, in my view, the same thing as negative growth. That expression is an absurd oxymoron, but it is a clear indication of the extent to which we are dominated by the imaginary of growth" (Latouche, 2009, p. 8); (3) "Strictly speaking, we should be talking at the theoretical level of 'a-growth', in the sense in which we speak of 'a-theism', rather than 'de-growth'" (Latouche, 2009, p. 8).

So, from the extract above and from the entire text, it results that degrowth is in opposition with the type of standard growth, limitless, productivist and economist, and in addition unfriendly to the environment; it is not a comeback, because a negative rate would be "a disaster"; to reach this goal—"living better, working and consuming less"—we need a new political system, with a new philosophy—a "society of degrowth"; an "ecosocialism" would be the most adequate political system, as André Gorz (1991, p. 87) would later add. In short, degrowth is a slogan and a pamphlet against the old kind of growth. And what is the old kind? It is not just the growth designed by Keynes, with its thirty glorious years, but also the durable growth as it appears in the economic definition of the Brundtland Report. No matter which adjective we might use to "spruce" growth and development—durable, endogenic, popular, equitable, authentic, etc.—they remain, if not "subversive", at least "toxic". Latouche believes that only by making use of verbal diplomacy, we are not able to diminish the toxicity of the term and of the disseminated phenomenon. "The word 'development' is toxic", he decrees (Latouche, 2009, pp. 10–11), a suicidal mixture of a pleonasm and an oxymoron. "It is a pleonasm because, according to Rostow, development means 'self-sustaining growth'. And it is an oxymoron because development is neither sustainable nor self-sustaining". (Latouche, 2009, p. 10).

If the above sends us on foggy grounds, a good metaphor may be and is redemptive. Latouche invites us to envisage a river that overflows. Degrowth would amount to its return to the normal layer. This metamorphosis suggests that degrowth is a reaction to the lavishness and "barbarism" of a model that has become the "cancer of humankind" (Belpomme, 2007, as cited in Latouche, 2009, p. 20). Choosing between growth and sustainable growth would amount to choosing between barbarism and barbarism, or in equivalent terms, between ecological catastrophe and the subordination of economy to private profit (Stengers, 2015). We should avoid these scenarios as much as possible! So, where should we go? We should not walk towards a wall, but towards "serene de-growth"! Towards a utopia as a "source of hope and dreams" (Latouche, 2009, p. 32). But this is a concrete and rational utopia, "a political project" to propitiate man and nature, nature and economy, everybody with everybody. People should not force the limits and they should be satisfied to "be content to be stationary, long before necessity compels them to it.", as Mill (1885, p. 595) had put it a long time ago. However, now he prefaces the Introduction to Latouche's Farewell to Growth.

If these propositions have not yet clarified it, a plus of understanding may be offered by the dictionary of degrowth (D'Alisa et al., 2014). Because words talk, isn't it?! Here are some samples: economicist totalitarianism, progressist developmentalism, eschatological beliefs, compatible society, ecological democracy, tourism disease, holiday disease, uproot, planetary gardener, economic horror, phagocytosis society, etc. The use of such richness of notions is intended to fight liberal spirit as well as every economicist spirit. Everything that has destroyed the social web, namely the "wrong paradigm" based on efficiency, performance, excellence, short-term returns, profit above all, etc., must be abandoned while there's still enough time left. And it should be done by giving back to politics its dignity, and to people their opportunities to hope and dream!

Is it clear?! It is for us and we are not happy with what we have understood. We understood that degrowth is neither growth nor degrowth. It is neither a concept nor a theory. It isn't negative growth either. It is aimed at the North rather than the South, which it does not avoid, but rather it advises it. It is a slogan against economicist totalitarianism and against productivism. It is as green as it gets! There are as many Yeses as there are Nos. And there is also an army of people called to defend the slogan and to persuade that there are chances to reach the little light at the end of the tunnel, even though meanwhile we may also destroy the tunnel as well, as it is also harmful due to its origins!

5.3 Degrowth in Search of a Credible Fixture

The project of sustainable development found its reason to be in the elimination of "collateral damages" such as those related to the post-war type of economic growth; it is meant to correct the relationship between economy and the society and environment, so that future generations may be content with what they inherit.

The promoters of degrowth consider themselves authorised to ascertain that the objectives of the Brundtland project are not fulfilled. This ascertainment takes place at various levels. Strained scholars see catastrophes everywhere. More realist scholars are willing to see some pluses as well; however, they are rather discouraged by the many minuses, which are large and unjustified. To avoid looking shocking and nonsensical, the aficionados of the idea of degrowth have been looking for reliable reasons to make us believe them that a change of direction is not only possible, but rather necessary.

It is not just a handful of people who diagnose and give advice. Worthy of our attention is the fact that most such analysts are not economists. Or when they are, by their dilemmas and distrust in growth they offer their support to many and various biologists, geographers, historians, environmental activists, etc. Even more interestingly, such characters also propose the construction of new sciences, alternatives to the Economic science, which they claim is not or is no longer capable of properly doing its job.

What arguments can be used and what is the foundation of the need for a change of paradigm in favour of degrowth?

The version proposed by Herman Daly (1991, 2008) is the most credible, created using the able instruments of the economist. He is not concerned with just any kind of growth, but with the end of uneconomic growth. Roegen's, disciple does not hesitate to bring homage to classical economists who held economic growth at the centre of their preoccupations and thanks to whom the cumulated historical benefits brought progress for humankind. Strolling along the same avenue of ideas, he further believes that the growing opportunity cost associated with the ever scarcer natural capital will, at one point in time, limit macroeconomic growth. The "given moment" has just arrived! The past is the past, and it was ok. But now Growth has become "uneconomical". In terms of à la marge reasoning, marginal benefits diminish as we extend and we "physically dilate" by consuming resources that are rarer and rarer and more and more expensive. He declares himself perturbed that the rule of the moment for the cessation of growth in a company is not applied at the macro level as well. It would only be logical. And if we apply it, we should stop growing. What we currently do amounts to giving our consent for an uneconomic growth in which we add failures to the GDP, but we deduct successes! Daly mentions other limitations of growth, too. He admits it is better to be rich than poor, a common-sense truism, but an exaggerated wealth in regards to several limits, does not seem reasonable to him. He ascertains, on sound ground, and with respect to other economists, that going beyond sufficiency will determine the end of the correlation between subjective happiness and GDP. Consumerism, growth for its own sake, oversaturation and other "overs" will be the terms adopted by the scholars from "La decroissance" to utilize Daly's thoughts. Anyway, Daly is convinced that the limits of true growth have been reached, in substance and not in volume and that we should think of a balanced, "less ample" economy! And we should do it with a preparatory exercise in wakening, "repentance and conversion"! In other words, this is not degrowth, but growth pondered with respect to all sorts of limitations, aimed at reaching the "sufficiency threshold". This message is somewhat more composed. We see where we went wrong, we stop repeating our mistakes, but we go on! And we keep a cool head with respect to economic judgments and more. We should take care to live within our means, but without heated thoughts about catastrophes. This discourse is compatible with the project of sustainability as it evades the suicidal "liberticide" dogma about just any type of growth, however, sustainable it may be, as the scholars from "La decroissance" believe.

An analysis on the edge between agony and hope is offered by another activist on the issue of sustainable growth: the medic Jorgen Randers (2018) in his book *2052 A global prognosis of the following four decades*. His discourse continues that of Daly's, but its delivery is somewhat more pessimistic. Permanently concerned by the future of mankind on planet Earth, as he insistently confesses, Quesnay's successor with a propensity for economics, tries his hand at offering a prognostic on the economy all the way to 2052, hoping that he would be proven wrong! Randers faces us with five big problems, on the solving of which rests the chance of a new paradigm: (1) the end of capitalism; (2) the end of growth; (3) the end of slow democracy; (4) the end of harmony between generations; (5) the end of a stable climate. Five ends and no beginning! One cannot say this is optimism at its best. For each one of the five rubrics, Randers finds both flaws and reasons for hope. By and large, he reflects, we will not abandon capitalism! If we have new accounting rules for the establishment of GDP, if we are careful about the excrescences of financial capitalism, if we consider the Nordic model as a successful one and we make possible substitutions with respect to resources, we might not face big problems! Economic growth cannot be given up, but it cannot go on either, as long as the ecological footprint suffocates the planet! Excessive consumption and consumption-based growth cannot last either. The ascendance of collective rights before individual rights seems to him a reinvigorating sign. Democracy is good, but it has big swiftness issues; or, changes require rapidity. This is why a more authoritarian regime seems more promising. He finds the example of China inspirational. Regarding the harmony between generations, Randers discovers as many question marks as possible! Young people will be dissatisfied and will have a hard time dealing with pensions and national debt! Environmental problems will make conflicts more and more acute. A maximum threshold of an additional 2 °C in global temperature should not be infringed upon. The big issue is that the common voter will not be capable of perceiving the catastrophic threat for the environment and will elect the wrong people, the politician who will feed him empty words, not frightening truths!

It is noteworthy that not everything seems strange in the lines above, and the word degrowth is neither proposed nor used. We cannot find it in the quoted book either. The target is the kind of current growth which, if seriously amended, may stand a good chance. We can go forward if we eliminate the "but" that hinders the consistency of the main indicator's dynamic: "productivity will grow, but meet obstacles" (Randers, 2018, p. 93); "Production (GDP) will grow, but more and more slowly" (Randers, 2018, p. 96).

Much more alarmist proves to be Richard Heinberg (2018), a journalist and educator on the subject of energy, who likes to take by his side names such as Krugman, Daly, Rogers, Exter, and others like them in order to talk using the language of economy. He even demonstrates some perspicacity in this domain and uses it obstinately to point out the catastrophism of a time about which he wonders how it is that we can still bear. His book *End of growth* would prove interesting and re-readable if it was not so full of frights. From the start, he announces that "The central assertion of this book is both simple and startling: *Economic growth as we have known it is over and done with*" (Heinberg, 2018, p. 13, author's outlining). Who expects magic in economics? If one tailors one's expectations following such a course, disillusions are just as big. Heinberg's book is full of disillusions. Here are some of them: salvation plans in times of crisis are "bridges to nowhere"; demand has become bigger than supply because of the many and various challenges facing production; humankind is about to experience the peak of raw materials; the decline of nonhuman species is a higher risk than terrorism; the externalization of the production of energy-consuming goods is an insidious solution: some grow, others degrow; the path of technical progress is saturated; automation is a vicious cycle; Schumpeter has reasons to be sad—docile robots will hit against a wall; are specialization and the division of work a saving grace? No, because we do not even understand A. Smith! (who is made fun of with his pin factory); would international commerce be a solution? Even less, as people were happier before trading; sad is the road of the service sector, but so is the path of globalization. It is also recommended to understand that "Gross national happiness" can be found in Bhutan, but not in the USA. The latter has growth but not happiness; we cannot go on with generating money out of thin air, debts may be reduced globally, but it would be better to "reinvent money" (Heinberg, 2018, p. 263). And, very importantly, political decision-makers must take notes: *now* we should avoid the monetary-financial show and not postpone the day of settlement! Etc., etc.

We could go on. The sorrow of living on a dying planet would deepen with every other example. The author's conclusion could be summed up as follows: a critic in search of proof for debunking the myth of the end of growth would do well to pretend that Europe does not exist. The causes are visible for everybody, of which three are the most important—the exhaustion of resources, the proliferation of negative impacts on the environment, and financial failures. The author believes that it should not be a surprise for anyone that growth has come to an end. And by using such titles as "From the frightening theory to the even more frightening reality", the same Richard Heinberg does not make it easy for the adults and children of Europe and the world to sleep well. With such fears subsumed to the idea that Europe and the rest of the world are condemned in an eco-alarmist manner, that everything has limits and our limit is not far away, what sort of ideology are we defending?

The redemptive idea to avoid the scenario to which unrestrained growth leads is the one according to which degrowth would be called for and applicable by itself. We wouldn't need too much theorizing on the matter! Infinite growth for its own sake in a finite world, the threatening ecological collapse, poverty, and everything that abuses the recovery capacity of ecosystems call, on their own, for reflection and urgent measures. Everybody can understand that. Roegen, the Medwases, and all their followers have said it plainly and for a long time. However, time has passed, we have done nothing, and there is the risk of heading straight towards a wall. We wouldn't deserve such fate! A simple logic compels us to make changes. Even the snail understood this, as Latouche reminds us, quoting in turn Ivan Illich: "At that point, the problems of overgrowth begin to multiply geometrically, while the snail's biological capacity can be best extended arithmetically" (Illich, 1983, as cited in Latouche, 2009, p. 22). When plus becomes surplus, it is advisable to stop! It is not only the logic of the snail that calls for the disruption of the "infernal circle" of growth. The expansion of an ostentatious consumption driven by deceiving publicity, the logic of credit subjected to the logic of dictatorial capital—maximum profit, the loss of any sense of measure in every concept that starts with *over*: activity, development, abundance, consumption, indebtedness, accumulation, etc.. Therefore, with ***over***-measure in everything that defines man, economy, and nature—all of these, on their own, ask for a radical change. It is no accident, Latouche believes, that such matters occur in electoral campaigns. They are full of direct, easy-to-digest messages. And then, considering the fact that the market is saturated, that homo economicus has been ugly even since the time of the industrial revolution; that we have been living on credit for a very, very long time; that we consume from the patrimony; that the South has remained the milk cow of the North; that globalization, as is well known, brings evil rather than good; that the physical limits is a cold-hearted subject; that even Boulding (1966) criticized the cowboy economy back in 1966; that, finally, the balance sheet of social justice has been looking bad for a long time, and the ecological footprint looks even worse, if all these have been happening for a long time, what else are we waiting for? Should anyone come and make us aware by means of a straightforward talk? No, we have to see it by ourselves, otherwise there is a wall waiting for us! This subject is too hot to need tribunes to trumpet old and painful truths to us!

The conclusion seems obvious: sustainable development cannot be created as it still is development, meaning it consumes resources. It is only degrowth as a phenomenon that would correspond to the syntagma «vivre mieux avec moins»! It alone may solve the inherent contradiction dwelling in the very standard definition of durable development—the contradiction between needs and limits—natural, technical, organizational, etc. (Vivien et al., 2013). A stationarity à la Mill is not enough. We must go beyond this state, "plus loin" as Latouche (2005) will put it. The contradictions faced by durable development, conceptually and phenomenologically, are exacerbated if, once inside Abraham Maslow's Pyramid of needs, we ascertain that the value systems, especially the aesthetic ones, have not been included in the concept. This notion only had in view the Western paradigm (Saigal, 2008). So, if it fails to include the messages of the Coran, if it does not talk about lust and cosmic revelations, what good is durable growth (for whom is it useful)? Degrowth would suit them better!

In the same line of thought, regarding the relationship between sustainability and the spiritual area of human existence, Kumar (2017) writes as well. The way in which the three pillars of durable development, economy, environment and society, are served by a version of development with the attributes of sustainability would very much be contingent on spirituality. If there is no spirituality, there is also no sustainable development, because "Spirituality helps us in keeping our greed for material well-being and resources in check and SD can only take place when we use the resources for our needs and not for our greed" (Kumar, 2017, p. 131). From this context, it follows that sustainable development has problems on this level as well, and that's why it is not really sustainable! We need something else!

In search of something else, a rapidly developing chaos takes over concerns, positions, analyses, concepts, etc. While no encouraging facts have occurred since the Brundtland Report, the theory on the subject has given in to madness. This is what Bruno Villalba and Olivier Petit (2010) establish in their article *Développement durable et territoires: quelques changements sans rupture*. Then, while the concept becomes routine through excessive quotidian referencing, the querela between the Anglo-Saxons and the francophones on whether sustainability equals durability and on their relation to resilience goes on (Blühdorn, 2015; Daly, 2008; Dobson, 2008). It is also important to see the way they position themselves in relation to the traditional concept of development (Berr, 2006). Against the same background severely lacking certitudes and clearly defined notions, they militate for new sciences: Environmental economics, Ecological Economics, Eco-economy, etc. It is worth retaining that under the roof of the same terminology, there is no unity of notions. For instance, ecological economics was inspired in the beginning by Herman Daly's vision on noneconomic growth. Meanwhile, however, it was seized by the much more radical movement of the advocates of degrowth, especially the scholars grouped under the umbrella of the ecological economics and policy program coordinated by Joan Martinez Alier and Giorgos Kalis at the Universitat Autonoma de Barcelona. In other words, the search for a new science is on, one which cannot yet decide what it wants to convey and under what form it wants to present itself to the world! Furthermore, as the achievements take too long, an audit appears to be justified, as well as the questions about the status of the phenomenon. Is it still a political project or just a "horizon"? (Zaccai, 2011). And one more thing! Didn't durable development as a political project include from the start a subversive dimension, for confusion and façade, to hide real issues under a modern looking armour? Everything is possible! (Zaccai, 2002).

Thus, reasons to search for "something else" exist! If facts do speak and they say it cannot go on like this anymore, and if the scholars of economics or environmental activists or even famous entrepreneurs are convinced that the economy resets itself at lower and lower levels, then what are we waiting for? We cannot pin our hopes in capitalism, tells us, among others, Ignacy Sachs. The creator of the "ecodeveloppement" believes that the paradigm of the twentieth century has failed!

And then, if Heinberg's sun cannot save us either, and if the world is finite and sustainable development has let us down, what do we do? We devise scenarios preparing for "something else"; or we contract a scenario-manic condition!

5.4 Degrowth Between Yes and No

To oppose degrowth to growth, one needs solid arguments, an auspicious intellectual atmosphere and clever minds to impose a new idea. Do we have all of these? We do have some, but not really!

We do not lack famous names in this regard, quite the contrary. But when we think we can count on them all the way, something—a fracture in logic, a hypocritical thought or too large a dose of daydreaming—is puzzling us in forming firm convictions.

First of all, we think it is easy to re-evaluate the deep gap between the academic line, friendly to the logic to be found in books, articles, conferences, etc. about sustainable development and the line of common-places that so routinely engage with the concept that it leads to its total dilution, turning it into a leitmotif and mark for everything and every-body wishing to seem cool! However, what we mean to say is that such cracks are found within the analytical exercises of important proponents of degrowth. Within analyses, they call for reflection and inspire credibility. In proposals, they encourage the use of slogans that dilute the message and make it less credible. Who adheres to such slogans as: "Work less, gain less, live better"? People who hate assiduous and thorough work may create a universal party to support such ideals, but we do not consider this to be the objective of the degrowth theory. Or, to draw attention, one may recourse to a mobilizing imperative such as "Stop growth!" This is something just as damaging for fostering serious and shareable beliefs as when one commands *End this depression now!* (Krugman, 2013). Both "Stop" and "Now" are irrelevant in the dynamic of the economy, and we do not know whether by making recourse to a shocking and opposing language one will gain the things that one intends to. Actually, such slogans are meant to make noise in the domain of scientific analyses, as well.

Secondly, there is a kind of awkwardness, more or less visible, in the behaviour of those who argue for degrowth, determined by the social and economic position of the platform from which the suggestions are made. For instance, the net division of the proposals for the North and for the South does not work all the way. Sharing the same world, the rich and the poor have too many existential interferences. We claim we send them different messages, but they are not really different. A message like "stop growth" is totally crude and callous for Africa! All the supporters of degrowth are clear in saying that "exiting the material" is not valid for the South. It is not just because the South has not even entered the "material", but because of the hypocrisy of this theory, we believe, is more fond of the degrowth of the wealthy than of fostering growth for the poor. If we consider the percentage of people populating the underdeveloped South, one should compulsorily be embarrassed by the wagons of energy and waste paper consumed on the subject of degrowth.

Thirdly, the confusion, the rope dancing between Yes and No, namely between hesitating and accepting either growth or degrowth, is caused by the conceptual intoxication that accompanies this phenomenon. If with respect to growth and the related notions, things are relatively clear, the same thing does not apply for degrowth. Latouche and his companions explain the unhappy situation of not finding a proper equivalent in other languages for the "descent" in Roegen's discourse. And in France, where the subject is animatedly debated, "decroissance" was the term approved by Roegen himself as properly representing his ideas. With all the efforts spent to show that degrowth is neither rollback nor negative growth, this has remained the preponderant meaning with which this term is employed, by use and abuse of it! With or without the will of Latouche, Gorz, Illich, etc.!

There is something else that may impair the chance of trustfully accepting degrowth: the suspicious dilemmas of some important supporters! They recommend "repentance": that is we have been wrong in promoting a "failed paradigm"! How far can regrets and forgiving go? The "honey" of growth is too tasty and after having smeared your fingers with it, it is hard to forget. So, Latouche does not forget how good the steak is, not from the South, but from the North. Let's hear him: "The values we need (altruism, conviviality, respect for nature) are also the values that will allow us to begin a dialogue with other cultures without destroying them in the same way as the arrogant universalism of a dominant power, because we agree to recognize the relativity of our own beliefs. I do not, however, make this an absolute principle, and I do not feel that I have the right to prevent a Hindu from regarding killing a cow as murder, not that that will prevent me from enjoying a good steak" (Latouche, 2009, pp. 102–103). Herman Daly (1991), as we have already shown, only wants a physical contraction of the economy. However, he also craves for the times of expansion because the benefits of modern life that growth has allowed for are always preferable to the precarious lifestyle of humanity's beginnings. We do not want to freeze either. We want to benefit from the comfort of a solid and warm dwelling, equipped with everything we may need. That is, we want growth! Both in the North and in the South. By such growth, we have access to steaks and a warm house.

We should also say a word about the well-known pioneer of degrowth: Nicolae Georgescu Roegen. The scholar's honesty is well-known. What we also know is that our compatriot left in 1948 a country headed straight towards the wall of communist starvation. At that time, 80% of the population of the country was living in villages. It was going through a rough time, needing growth like they needed air. An outdated system cuts its wings. This country needs growth today as well; real growth, not one fostered by consumption. It is not a poor country, but even today, making it degrow seems almost absurd.

The indecisive attitude towards degrowth is explained by the fact that, at least apparently, it is hard to draw the line and conclude which argument should prevail within the analysis. We should rather say that none of the parties *seems* to have the force of annulling the other. Where does this "seems" come from?

5.4.1 From the Puzzling Suggestion that Degrowth Is An Invitation to Stop by Continuing to Run

Firstly, we want degrowth, we want to stop, but at the same time we want to go "plus loin", beyond capitalism and to nowhere! Conventional growth, with all its shortcomings, is associated with capitalism. It is clear that degrowth must do something about it! Its accommodation and reformation are subject to criticisms and are actually criticized. J. M. Harribei and A. Gorz are very careful with the "purity" of this phenomenon. Criticism, they believe, is no longer needed. Marx did it and, according to Latouche's point of view, if ecological constraints had been included, Marx's criticism would have been excellent. Anyway, this is not the time for criticism! It is time for a "leap forward"! «Plus loin»! Or at least an ecological cultural revolution. "Our conception of the de-growth society means neither an impossible return to the past nor a compromise with capitalism. It means going beyond modernity" (Latouche, 2009, p. 90), he clarifies what he means by "plus loin". Even though the word *going beyond* is marked with inverted commas, in both the text and the context it is intended to mean a movement, even a leap. This should be a "commanded" leap, as Latouche and his companions are aware that capitalism would never make by itself the decision to limit its growth. And if it cannot be stopped by an ingenious manoeuver, it can be outrun anyway! A "commanded" manoeuver is also necessary because the "enemy" is firmly grounded.

Secondly, by calling for degrowth and demythization, we invite man to take part in an exercise against his nature; we invite him to stand still. We fight against the myth of growth! Who hasn't heard of it and who hasn't believed it? A victory against it amounts to the instauration of another myth—the myth of degrowth. To impose it you need time, patience, force of persuasion and validating experiences, plus attractiveness, sympathetic capacity of the "slogan" leaders to coagulate adherents to follow and support them. What percentage of the world today is interested in degrowth and in how many agoras can one hold a discourse on this matter without being suspicious? Ecological issues impose a growing trend. Before becoming representative, the idea as such needs more arguments, not just an invitation to taking leaps! When we talk about the fact that it is movement, not stillstand or return, which naturally defines the dynamic of the economy, every economist can find generous support. It would have been easier for Latouche to refer us to one of the first interpreters of economic thought, his conational Antoine de Montchrestien, to find out that "man is born to live in permanent exercise and occupation" (Montchrestien, 1889, as cited in Denis, 2016, p. 105). It is "exercise" and movement rather than stillness in time that defines man. Montchrestien talks about an "exercise", a movement that accompanies work, the only known source of wealth! Instead of such sources, the author of *Farewell to Growth* refers us to the well-known critic of what is to be "outrun"—Karl Marx. However, we should not try to fool people,

Marx himself hated stillness. In his *Fragment on machines*, he clearly explains how capital enslaves labour through humiliation, subjecting it to an exercise commanded by the master of capital for the purpose of facilitating the effort and increasing productivity. Once technical progress is implemented, you would expect Marx to promote statism and deceleration. But he doesn't do it on principle. Work is not substituted by the "Laziness" of his son-in-law Lafargue or by the work games à la Fourier (1846). No, work is still work. It will escape the temporary humiliation of capitalism so that, to use his and his friend's Engels language, it will elevate the human being once communism takes over. And "elevate" it did! Marx admittedly also realized that the "enemy" cannot be stopped. Revolution was the solution he proposed. In contrast, degrowth does not want a revolution, but a huge effort of making people aware through an orderly march past modernity.

Thirdly, the economy may only stand still in theory, for the purpose of taking a picture and studying it in school. It is well known that the real economy does not stop: it either grows or contracts. Static equilibrium and optimum are didactic exercises necessary to speculate on the impossible and the ideal. The economy is always in motion! Walt Rostow's (1990) or Simon Kuznets' (1971, 2013) staged movements are famous. The latter shows the evolution of the modern world towards the third sector, the victory and the ascendance of services in relation to the "material"—extractive and productive branches. It is a large movement that defines the level on which technical progress succeeds in bringing about civilization. In the first two stages, speed seems perceptible and measurable with the naked eye. The countryman of the first stage moves slowly. The worker of the second stage moves fast. The worker in the service sector "cannot be seen"! We may assess his speed by using an average, but we do not possess the real measure. This is why a phantasm is also possible! It is possible that the hardly imaginable imagination of those who call for the dematerialization of imagination is also subject to an optical illusion. Once we reach the end, we move in place; only the humming bird can compete against us. We stop, but we move in all directions, and nobody can detect our doctrine!

5.4.2 *From the Contrast Between the Results Promised by Degrowth and the Utopian Character of the Project*

Finding the demarcation line, between science and nonscience, has been a difficult problem to solve for Economic epistemology. Degrowth can be consistently nurtured by the battle of those resolute in imposing it against those more hesitant to do so, but this does not add to its chances of clarifying its old obsession (problems). Probably, only Thomas Kuhn may help us distinguish the things that are worth constituting into a hard core from the whole lot of contradictory arguments so that, step by step, degrowth would become a paradigm with scientific attributes and major recognition. Until then, we keep swimming between Yes and No.

Degrowth draws attention and invites not only to meditation, but also calls for a "political project" to address the big issues, both old and new, presented under the guise of catastrophic attitudes concerning the environment, resources, population, disparities and social justice. When harmony with nature is broken, the ecological footprint exhibits unsettling signs, the prognosis of non-renewable resources cannot promise a secured future, population grows too much where living conditions are not sustainable, when the disparities between the privileged and the poor are not encouraging, when social justice remains only a mirage, when too many people live badly etc., etc., it is clear that the degrowth "slogan" claims its right to a Yes, without doubts or hindrances. And so? Where does the dilemma of breaking with a system and then embracing another, comes from? It comes from the huge difficulty in leaving something that is not altogether worth leaving. The artisans of degrowth do not hesitate to declare that degrowth is incompatible with capitalism and that they want "something else". However, this something else is not yet definable as long as they only claim to cut the umbilical cord (those with recognized socialist-Marxist orientation really do it!) with a system which has truly revolutionized, as nothing else did, not only the economy, but through it, life itself. They just make such claims, but they do not go all the way. And they don't because there are counterweights and incontestable reasons that do not allow, not even in theory, a break. Not even in regards to the outlining and measurement of the ecological footprint, they cannot break with capitalism and its defining corollaries—accumulation, capitalization, profit and growth. Why? Because without the growth, there would have been no technical-scientific revolution. Knowledge and the knowledge society would have remained a utopia. It is only with the support of ultramodern technology obtained thanks to growth, that we are able to diagnose with the intended precision the seriousness of phenomena pertaining to the environment. How many satellites does Africa send into space to predict the weather? This circle is not vicious, but rather logical; if there is no growth, there is no technology and no knowledge; and without these we have no chance to find out where we are.

Growth has given us shoes, offered us shelter and fed us; at least to some of us. Others will follow. How can one deny it? Nobody does. The issue is one of proportion: how many pairs of shoes, houses, dishes, etc. do we need to feel wealthy within reasonable limits. From this point of view, we award a "Yes" to Latouche's argumentation. But there is also a "No" coming. It is a "No" awarded because of the notion that "acceptable needs should be determined by the community as a whole" (Latouche, 2009, p. 54). A communist experience might forewarn anyone about the hideous features involved in the implementation of such a notion. When you have your menu already established by a party or some "municipality", you no longer live in a free world. Degrowth wants to go beyond capitalism, skipping it the same way a runner jumps over an obstacle! Up to this point, things are clear. However, "the elimination of capitalists, the ban on the private property of production goods, the abolition of wage relations or currency would plunge society into chaos" (Latouche, 2005). So, the intention is not to throw away capitalism altogether! We do something nobody has ever done before: "To escape from development, economics, growth does not imply giving up all the social institutions that the economy has annexed (currency, markets, even the wage relation) but it means to re-embed them in a different logic" (Latouche, 2005). It is a mishmash, a creature resistant to the weather, one that can be milked while also being capable of running fast when the floods come! In this instance, André Gorz was more limpid. His "ecosocialism" is "the positive answer to the disintegration of social relations under the action of market relations and competition, specific for socialism" (Gorz, 1991, p. 87).

On the whole, do we or do we not get rid of capitalism with its turbulent growth? Displaying it in a new reliquary may provide the solution: a living dead! If you had to deal with a pigmy, it would be easier to promise your supporters a victory. But you actually want to shake a system rooted deeply not only in the soil of the economy, but also in the collective mentality, in the philosophy of life and politics, a rational system that grants a shot at prosperity. It has many excrescences. It is owned by multinational companies. It's not easy to get them out of the way. All the supporters of degrowth know and say that. But this is not the only difficult part. It is hard because the global economy is complicated and cannot be mastered using one recipe only, be it even degrowth. The South has its issues, but it is connected to the North, and vice versa. Resources are distributed very unequally by nature. The initiative "think global, act local" has to deal with the specificity of each state. The idea may be good, but its implementation is utopian. Utopian is also the idea that every individual will have to escape productivism and economism so that the reduction of GDP is a resultant, not a starting point. Using a normative act or a command would not look good! Willingly renouncing an excess in one's living conditions may sound like a reasonable idea in developed countries, but one cannot say it is also achievable. For underdeveloped countries, the idea does not exist.

And so, what can we say? Yes, for the temerity! No, to the utopia. It's suspicious to forget that a healthy and durable growth creates jobs, that it is desired by both the rich and the poor, admittedly to different extents, that growth is better than redistribution and that this is the only chance to achieve a plus of justice. It is hypocritical to pitch degrowth to the poor, just as it's disconcerting to propose this recipe to a country with large foreign debts.

What the supporters of degrowth see today when looking at an economy that works, but does not check the objectives of durability, is an "infernal machine" which, fuelled to the extreme, emits smoke, intoxicates the atmosphere and abuses resources, supporting an ostentatious and irrational consumption. An outsider of this movement such as Wallerstein considers that even accumulation has become irrational. He acknowledges that for 500 years, "The modern world-system has lasted some 500 years, and in terms of its guiding principle of the endless accumulation of capital it has been extremely successful. However, as we shall argue, the period of its ability to continue to operate on this basis has now come to an end". (Wallerstein et al., 2013, p. 11). A hegemonic capitalist system, unstable and generator of disorder, must be fined! However, as a saddened Wallerstein observes, there is no one left to revolt; there is no working class left to do this, just an inert mass of clerks, intoxicated by the affliction of a myth that has absorbed and integrated them. Hence, the need for a detox, mentioned by Latouche, and the need to réenchasser or to "reinsert", to use Polany's words.

Let us finish this journey between Yes and No with Jorgen Randers (2018). To the question from his subtitle *The end of economic growth?* From page 43 of his book, he answers with a claim and a reformulation of the mentioned question. The claim: "I believe that society will continue its effort to seek continued economic growth, among other reasons because this is the best-known method to create more jobs and facilitate redistribution. (…) Thus the proper question in my mind is the following: Will humanity manage to limit its ecological footprint to fit within the carrying capacity of the planet?" (Randers, 2018, p. 43). We consider this position to be a credible, defining and encompassing finale to the never-ending duel of ideas between the pros and cons on the subject of degrowth.

5.4.3 Because of the Lack of Coherence in the Remedies for the Developing World

The world of the South, mainly Africa, is not just full of Yes and No; it is also filled with paradoxes. With a complex and complicated cultural matrix, a different colonial legacy, different stages of underdevelopment, serious issues with respect to food, population, injustice, corruption, healthcare, security and integration in the world, this is the economic and social reality least suitable to the standard discourse on degrowth. The invasion of productivism, the industrial or agrarian revolution and material imagination, as well as the imaginary material, all of them avoid it. Another lesson than the one on degrowth would befit it. However, the South is not exempt from this kind of analysis. This world that represents a large portion of the global population is analysed by many scholars and from all doctrinal perspectives. And while the "pieces of advice" prosper, the successful examples do not gain representative influence, quite the contrary. The prophets of degrowth do not miss the chance to discuss the subject. However, they fill it with inadvertencies that are hard or impossible to use when constructing syllogisms bearing the footprint of this world, which is, in itself, the culmination of everything that may be imagined as pure paradox.

The "promotion of degrowth for the South" fills 11 of the 153 pages of Latouche's *Treatise*. They are written with passion and are rather disturbing! They are as full of exhortations to reflection as they are vexing and nebulous. A proven and excellent connoisseur of the realities of Africa (Latouche, 1998), he commands trust. It's just that this would be too nice to be true. He says in a clear and imperative manner that "reducing Africa's ecological footprint (and GDP) is neither necessary nor desirable" (Latouche, 2009, p. 57). It's logical; one cannot return without leaving first. However, what is clear in the first instance becomes muddy in the second one. Degrowth does not seem to evade this world, on the contrary, it looks at it directly inasmuch as "Reducing Africa's ecological footprint (and GDP) is neither necessary nor desirable" (Latouche, 2009, p. 57). However, this adventure seems to have already been over. Under the guise of "participative" or "popular and solidary" growth under the banner of socialism, liberalism, or simply capitalism, sustainable development has proved, Latouche believes, especially in this instance, an "uprootal", a "mystifying anti-economy" and a "monstrosity" that ended in corruption and poverty. And so, what can be done? A change of direction seems to be the solution. However, the problem lies with the direction itself. With the help of other "R", five in number (Rift, Resume, Retrieve, Reintroduce and Recuperate), he basically suggests a break with the North and an attempt to reconquer their own past and retrieve a tradition that once proved generous in securing food, a quiet life and even happiness. Everybody knows that the South is entitled to reconquer its own past of which it was painfully and irrecoverably castrated by that part of the North, animated by colonial uneasiness. The problem is how? And here, the supporters of degrowth fill the discourse with reveries and dilemmas.

Angry because in front of him laid an Africa full of second-hand products, "Western rubbish", Latouche (2009) exhorts the separation from the West and the "farce of durable development". He has, we are convinced, reasons to be angry in his attempt of redrawing a perimeter and a separating fence, a way of escaping the contagion of this "disease". This should be a perimeter within which the recommendation would be to "stay afloat": that is, far from the "intellectual and political elite of the continent", opposed to "Western engineering, industrial, and entrepreneurial economic rationalities"; using one's own currency and finding support in neo-tribal connections! Our author wants something, but he knows that what he wants is not possible. He wants, first of all, to break with the world! He proves clumsily in writing that "dc-growth in the North is a precondition for the success of any form of alternative in the South" (Latouche, 2009, p. 58). The "triumphant and arrogant" globalization does not elude the South. And, interestingly, it is successful among the young. They are aware that it's worth refusing the "inferno" back home for the brightness of the North, "against whose gates they will crash". This is true, but no one can prevent them. The two worlds call each other like two magnets, and the people who make the junction are predominantly the young, who are not satisfied with their condition of being "excluded from economic modernity". How can they be mesmerized by a discourse about getting by using one's own means and a return to the "self-production of the countryside"? The "absurdity of mimetics" is to be accepted. However, the North's mirror is too visible and attractive, no matter how artificial might be the paradise that sits behind it. The relationship between generations deserts the traditionalist model, even in the case of an African continent rich in cultural and religious motifs, in brotherhood with nature and a profound humanity. Latouche risks leaving unfinished his project of reconversion if he were to rely only on the seniors. Actually, he even confesses that Albert Tévoedjrè (1982), by whose book he is inspired, a militant and also a politician holding state offices, could not implement his convictions. Why not? Because, however, parental, enrooted and compensating his home culture may be, it cannot rival the Western one. This is the demand-and-supply relationship, and nobody is responsible if one prevails against the other. However, Latouche, in anger, refuses that which cannot be refused. He refuses or dilutes the role of the elite and of the "engineering and entrepreneurial" economic rationale. He makes a habit out of only quoting Mill when he needs to lean on him in order to convince that a break is in order. Why does Mill fail to "bother" him regarding the radical role of the elite in the making of a society? Why does he fail to recommend the same author's book *Considerations on Representative Government* when he visits Africa? Actually, the notion that one could get rid of poverty and misery, be it even in Africa, without engineers, economic rationality and entrepreneurial spirit, nobody can believe!

"One has to choose between these two options", challenges the great believer in the virtues of degrowth! We either help them and follow the "alter-mondialists"—an allusion to J. M. Haribey—or we leave them alone! Help, Latouche admits, is "suspicious". This is true, but how is it then possible for the North to repay some of its debts? In this context, a speculation takes shape; Southerners have no clue what they want and, intoxicated by the consumerist mythology of the North, are willing to ask for mobile phones and cars instead of food! That may be so! A natural and good thing to do would be to "teach them how to fish", to introduce them to the craft of production, not to deliver them the "fish" already caught. This message is included in the Brundtland Report, but Latouche and his companions cannot convey it: it belongs to the philosophy of sustainable development, from which Africa itself should run! However, leaving them alone doesn't work either! According to the "concrete and fertile utopia", the poor and the marginalized of the world must be left in peace to deal with their own "self-organization of getting by". However, at the same time, they must be taught how to avoid the pains, errors and excesses of the North. Caught in the net of his own contradictions, Latouche avoids giving direct advice. He delegates Pierre Gevaert (2006) to do it, with his *Alerte aux vivants et à ceux qui veulent le rester: pour une renaissance agraire*. With respect to the world of the African continent, the absurdity of the practical prescriptions on degrowth reveals its full splendour. Convinced that he must say something on this subject, and lacking substance, but compensating with a lack of clarity, Latouche preaches: we either help the Africans by letting them manage on their own, or we leave them alone, but we tell them how to live. The French reveries on degrowth cannot help Africa, but the growth recipes of the founders may provide people with a chance for a better life. In this case, there is no choice, as for the South, degrowth is not an alternative.

"What about China?", Latouche wonders in a "ritualic" manner, in relation to degrowth. On the whole, he believes, it would benefit it, especially when the world's fortune depends on its decisions. He does not hesitate to indirectly give the Chinese a piece of advice: "a resolute commitment to a de-growth society that demonstrates that the 'model' is desirable and therefore exemplary is the best way to convince China, India and Brazil to change direction"! (Latouche, 2009, p. 65). What are we to believe? Let us see what others have to say about that. Alain Peyrefitte (1976)? Should we understand that, if mimetics is not recommended for all countries, then China is the best example for the veracity of this notion, and that its totally specific realities does not tempt it to borrow or to offer a model? Should we recommend it for degrowth? It already has too many mistakes to learn from and does not need anyone else's. And we also think it doesn't trust Alexander Gerschenkron (1962) either; it is not willing to believe that the disparity may be a stimulus and an opportunity to see what others did wrong so as to avoid their failures. It is too rushed to use all its engines to recover and get in front. It is interested in the ecological footprint because it has no choice, but it is in no way interested in degrowth.

We do not think we can blame anyone that, when faced with a problem such as the one of the poor world, he gets bogged down in dilemmas and, because of this, he becomes less credible. The real issue lies elsewhere. The fragile credibility is due to the stubbornness with which, on behalf and to the claimed benefit of this world, a time and again verified paradigm is refused in favour of another one, full of well-known utopias and politically impossible to implement. Of all, the refusal of the important role of the elite as well as entrepreneurship in the countries of the South is a lacuna that is difficult to swallow. This is because, as the founders have told us, without them we cannot overcome the impasse and we are going nowhere. First, getting out of poverty presupposes making people aware, not just individuals—which is more plausible to happen—but also the masses about what it is and what is to be done. The admission of the current state of affairs, including our ignorance, is salutary and appreciated. But how does it happen? It is like the disease of dogmatic people; when they become aware of it, they get rid of it. It is necessary for someone to "ring the bell"! And this is the role of an unsubordinated and uncorrupted elite. It and only it may dictate the agenda of the start. And this agenda essentially amounts to a hierarchy of priorities. It is important to know the things to start with and what piece of the ensemble carries the others. If they should start with rules, money, market, property rights, education, the open attitude towards the world, the employment of women or the fight against malaria, this is best known by the political-administrative elite of each country. If a country does not have one, the usable example of the "Chicago-boys"—the scholars that successfully kick-started the economies of Latin America—may be a good idea. To learn by doing, but doing, not standing, without necessarily taking anyone else's example, has proved to be a good way to act. David Landes, an author who outlines the sources of possible economic development, as valid for the North as they are for the South, says that "all the old advantages—resources, wealth, power—were devalued, and the mind established over matter. Henceforth the future lay open to all those with the character, the hands, and the brains" (Landes, 1998, p. 285). This is the triad of the new paradigm: character-arms-brains! Are all these to be found in *L'Autre Afrique*? Yes, enough and to spare. They are just waiting to be refined and to come into contact with the world of knowledge. And for this, the youth of Africa, and not only they, must be allowed to enter this world. Only after having been schooled like this, when they get back "home", they may bring about development. They would be less interested in the intricate language of their project. But they will most certainly be interested in a project rather than a utopia, be it a "serene" one!

In other words, what we mean is that degrowth would only have to gain if it renounced not just the movements of right-about-face, which sabotages its logic, but also the stubbornness with which it claims the unicity of the same logic. If we leave aside the ecological footprint, which summarizes the direction of the future, all the other circumstances call for specificity, for both those supposedly standing still and the ones going back. The time of an imposed culture is long gone. On the edge, by choosing death, the Tasmanians collectively refused a foreign culture. The "Arabian springs" seem to also be examples of such refusal. Culture changes China. But it is its own culture, on which it alone can be grafted. A borrowed or poorly implemented culture may prove to be a hindrance. Weber explained it well.

5.5 The Scenario-Manic

We use this title, contaminated by the language used by the supporters of degrowth to greet us! For a game with the future, scenario-manic is the best adjective.

Since 2007, when he published Farewell to Growth, it is likely that to the eight "R"—Revaluation, Reconceptualization, Restructuring, Redistribution, Re-localization, Reduction, Reusage and Recycling—that define his scenario of "serene, convivial, and sustainable degrowth", Latouche may have added further "Rs". Even so, we do not think it would change the essence. It amounts to "let us be satisfied with staying where we stand". This advice belongs to Mill, and it is used to preface, as we have shown, Latouche's opus.

Neither Latouche nor those that he quotes and with whom he builds the scenario of leaving capitalism and economism, deny that two great minds lay at the origin of their intentions: Malthus and his updated version, Nicolae Georgescu Roegen.

The entire scenario of the new paradigm of degrowth is built on the sad philosophy of the Romanian economist. His fatalist spirit as well as the solutions it has inspired follow, as a shadow, the apprentices of degrowth. There are more than a few of Roegen's sentences that have passed, like an arch over time, directly into the mind and pen of the promoters of degrowth. For instance, the notion that the first law of thermodynamics is no longer able to serve economic logic is treated as the "letter of the law". The pendulum movement, demand–supply, and the formation of a "correct price" on the free market, "tamed" to substitute everything with anything is no longer inspiring. The economic process is entropic, Roegen says, and it is only the second law of thermodynamics that may be used for the rationality of a mechanism that consumes and, in a one-way manner, transforms everything into "garbage". The *Entropy Law and the Economic Process* also transmits further messages for the configuration of the current scenario of degrowth. The idea of a system, according to which the economy does not operate in isolation, bereft of social, political, moral and environmental factors, is an example in this respect. Thus, the difficulty of changing anything determines the denial of the *entire* system of today; their scenario-manic leads to another society, it's not just a new piece in the old mechanism.

Then, a further big idea comes along, as a lesson that the poor may teach to the rich. The Mengerian gradation table of needs fails to tell us what Roegen's poor people can say, namely "that of all necessaries for life only the purely biological ones are absolutely indispensable for survival" (Roegen, 1971, p. 277). This is an impulse à la Latouche or Mill, meaning that it is satisfaction rather than ostentatious consumption which gathers support from such a point of view! It's not by tripling the number of personal automobiles that we become happier; and it's also not through marketing that "packs" the fundamental needs in the packaging of false, futile and wasteful needs. It's just that, Roegen warns us, there is something, both then and nowadays, that takes us away from this lesson; it comes and confiscates our smattering of rationality and makes us prisoners of the "pleasure of living". And with this occasion, we receive the lesson of the day: one of limitless acquisitions, ostentatious consumption and promotion of our own personal "brand", while forgetting or leaving aside the fundamental needs. This occurs by means of marketing that claims to make this into a science, convinced that demand is condemned to fall behind supply, and thus it must be "assisted". In this space occurs the struggle against "economic imagination". Efforts are made to persuade that you do not have to reach or get beyond satiety to feel well. Also here, a hopeless struggle is fought against the great "enemy"—the pleasure of living! And against the pleasure of living better and better, no proper weapons have yet been invented. Communism anathematized consumerism against the background of the "Law of the growth of living standards". After almost half a century of experimenting, it has not succeeded in defeating the "pleasure of living". From this point of view, maybe the only one, there is no "path dependency". As soon as they came out of communism, people "rushed" to the supermarkets! And they desperately engage in an exercise of recovery. This is an "R" that Latouche failed to include in his scenario-manic. And there is no way he could do that, as it comes from another logic, one that is imposed without *Treatises*. Degrowth "Bibles" nowadays validate the hypothesis that the price for the "pleasure of living" is to be found in the entropic march of the natural environment whose end of the road—pollution and wastefulness—compel scholars to speak in Roegen's language—as this classicist of degrowth prophesized.

In his language, Roegen built a little scenario as well, including what needs to get done and how things should work. Malthus's shadow does not leave him be, and that is why his first point concerns population. Both Malthus and Marshall seem exemplary to him. The great neoclassicist drew his attention that it is biology rather than mechanics that is the "true Mecca" of the economist. It's just that the economist, Roegen believes, remained as "stubborn" as the "client" from whose head he had to extract the fixation on the "pleasure of living" and who failed to see the irreversibility of the supply curve as a fatality! And he didn't see or he pretended not to see that everything happens within a given, well-marked, and unextendible framework. Thus, the entropic dance of the world is a zero sum game. In such a context, one is forced to adopt global and holistic solutions. This is the case with the population. One is compelled to relate the population to the resources of the planet. However, it's not actually the number that counts. The real problem is not to figure out the number of people that can be fed at a certain moment in time on planet Earth. Rather, one must figure out "how long can the earth maintain a population of ..." (Roegen, 1971, p. 20). And this is not enough! We must know what happens "afterwards"! Because there is no infinity to work with; because it is important to know what is "the maximum of life quantity that can be supported by man's natural dowry until its complete exhaustion" (Roegen, 1971, p. 20). The end, it's obvious, is unavoidable! To push it farther and farther, generations must collaborate, and in determining the population optimum W. Petty is no help, but rather organic agriculture. It is the one that is fed with its own products, refusing substitution; "the bull and the manure" versus the tractor and chemical fertilizers. Organic agriculture is the mark and the lever responsible with indicating the population optimum. Roegen comes to the notion that in the perimeter of the "minimal bioeconomic program", humankind must reduce its population to the level at which it can be fed using organic agriculture. Not one more!

Basically, Roegen does not intend to treat things from the position of the fatalist, but this is what he achieves. A radical change in the technical mode of production by means of a "Prometheus III" moment would open the opportunities for an energetic reconversion that would free us from the dependency on exhaustible resources and open our path to the sun! He is not very enthusiastic that we may soon get there, and he subtly hints that it's not even controlled fission and solar energy that can address a limitless world. For now, struggling with incertitude, Roegen believes it is proper for us "to be worried"! Let us think that the iron for a tank or a Cadillac might be used for a plough, and that it's not "gigantesque things" that make us "priests of ecology". And finally, "a systematic conservation" of our inherited legacy must be the axis of our conduct.

The market? To Roegen, it does not have any significant role. It only involves current people into the entropic dance. This is a shortcoming that invites state authority to regulate the consumption of resources to also favour "our future fellow people" as well. A summary of Roegen's "bioeconomic minimal program", with the interdiction of war production, the necessary support for underdeveloped countries, the avoidance of overheating, overcooling by use of regulations and the reduction of population in relation to the potential of organic agriculture etc. is accurately captured by Michalas Wade (1975) in his work, *Nicholas Georgescu Roegen: Entropy the Measure of Economic Man*.

What do today's fatalists have to add to this? They say that nothing or almost nothing works well. Resources are almost finished, poverty leads to extreme misery, the stress of the population is at its height, the evolution towards the service sector is unprofitable in regards to productivity, substitutions become more and more difficult, the perspective of people's uselessness looms on the horizon, globalization annihilates national states, marketing compels us to buy stuff we do not need, world peace is almost gone, while multinational companies and banks erroneously orchestrate the game of world economy. In one word, Jared Diamond's (2014) collapse is happening right now!. If, in addition, the ecological footprint is no longer kept in check, there is no doubt that everything must go through the n + 1 "Rs" of Latouche. An inventory of the references on which he bases his assertions are enough to see whether or not it is worth getting up the next morning! Here are some of them: *Degrowth or barbarism* (P. Aries); *Before it is too late* (Dominique Belpomme); *How can you become progressive ... without becoming reactionary* (Besset Jean-Claude); *La terra e finita* (Bevilacqua Piero); *Alertes santé* (André Cicolella and Dorothée Benoît-Browaeys); *Pour un catastrophisme eclaire* (Jean-Pierre Dupuis); *Le malaise americain* (Paul Hazard); *Unde societé sans école* (Ivan Illich); *Comment les riches detruisent la Planete* (Kempf Herve); *Eloge du luxe. De l'utilite de l'inutile* (Paquot Thierry); *Quand la misere chasse la pauvrete* (Rahnema Majid); *La fin du travail* (Rifkin Jeremy); *Welcome to the age of less* (Tomkins Richard); *Le viol de l'imaginaire* (Traore Aminata); etc., etc. Catastrophes and clogged roads!

How does one bring this "calamitous" evolution to an end?

Is it by controlling population growth? This is a "lazy" and perverted solution! Just as Roegen thought that the problem is not the number of population, but the capacity of organic agriculture, all "Latouchians" consider the solution to be a more just distribution of resources and results! What about the environment? We share it with other living beings, "leaving them, for example, 20% of the terrestrial area that humankind has not appropriate yet" (Besset, 2005, p. 318). With respect to what is left for us humans, our great "concern" should be the ecological footprint. The reduction of "intermediary consumption", "green chemistry", "the internalization of external negative externalities", clean production, the reduction of useless consumption, and generally, economic degrowth are handy solutions. What about the nation and the state? Ideas may flow unhindered, but material values have boundaries! The territorial anchor, considered all the way to local and "sublocal" levels, constitutes a vibrant splash of colour for this scenario. What about the rest? By law we may stimulate the recovery and the transformation of cities into centres for the processing of recovered metals. On small areas, one may remake the traditional social networks according to the "gastronomical" recipe of Ivan Illich (2003). Yves Cochet's (2006) "durable, biological peasant agriculture" is a replica of Roegen's "organic agriculture". Local monetary policies, not like those of Hayek but rather like the ones of Bernard Lietaer (2005), are welcomed. Or, as we have already learned, it would not hurt to "reinvent currency"! Fewer supermarkets and more mini-markets, a nutrition "less energy-consuming", the orientation towards a job-creating ecological economy, etc. look like common-sense and rational measures. The issue does not concern the opportunity and the rationality of these measures, but rather their implementation. Why so? Because the intention is to get out of the system. They want to escape the "society of work" and capitalism. They want to escape economism, productivism and the "capitalist logic". They want to secure the victory of Andre Gorz (1991) against Adam Smith in the struggle to rehabilitate the non-productive. They want to elicit such ulcers of capitalism that even Marx failed to see. And if we do not want to destroy the planet, what do we replace capitalism with? With an eco-socialism, neither right, nor left; anyway, a system against globalism, liberalism, free market, individualism and productivism. And if possible, the right to take a break, and even be lazy, should be guaranteed by constitution.

The socialist shell that encapsulates such proposals whose practical implementation is illusory even for their initiators, make the discourse convoluted. What good are the inciting analyses that give warnings and advocate for brakes against the slippages that the free market economy is known to go through? The packaging in which good ideas are sold ruin the entire "business". The scenarios of degrowth remain chimeras, with or without the will of their supporters.

5.6 Concluding Remarks: Between the Natural Rate and 2° Celsius!

The discourse pro degrowth is one of "opposition". If the artisan of the "serene utopia" became the chief of a government, he would be forced to realize that both degrowth and growth are intricate business, full of compromises and uncertain causal relations. From this point of view, the entire effort of supporting degrowth has the significance and the value of a warning. When you are in opposition, you do not praise the official policy; rather, you are searching for its flaws.

By using fan-shaped criticism on all levels of economic dynamic, the degrowth theory's inventory of truths is abundant. The fact that the Northern development model is successful is worthy of notice. The fact that the GDP must be subjected to a new accounting methodology, even though the warning is not new, is a justified criticism. The Genuine Progress Indicator, GPI, proposed by Daly should not be neglected. The fact that growth has become a problem rather than a solution, especially in the field of environment, is an incontestable truth. The demonstration of the "rebound effect", under actual circumstances, is obviously required. It is good to say directly that globalization amplified a lot of bad problems. The whopping lie of nominal growth and the endemic sin of growth through consumption are correctly emphasized. It's good to know that democracy is slower in adopting emergency decisions than an authoritarian government! That the young have to solve the problems of debts, and pensions is also good to know. That we may find shelter from the diluting effects of globalization in "local" areas, in "district republics" and in small communities is also good to know. That there is a maximum admissible threshold of global warming beyond which environmental turbulences put our life at risk is a must-know. How can one not know that social disparities and population problems are as serious as environmental issues? That we have left the vocation of durable things for the sake of insignificant things that master our lives is something that we must be warned about! That good and respected Law may stimulate the achievement of the eight "R" is a truth that must be accepted. The great firm may admittedly severely destroy the social fabric. We do know, but it is good to be reminded that we do not need 100 pairs of shoes to have an elegant and healthy walk. That those left behind may learn from the mistakes of the people that went forward is an almost biblical truth. That there is a problem of honour between North and South that needs to be solved is worth noting! That such large projects as the one on the conversion of solar energy require the participation of the state must be known. It should also be admitted that the city will grow, will get more and more complicated, and will attract the greatest part of the world's wealth and poverty, etc.

OK! So, what do we do now? Do we offer a standing ovation to the disciples of degrowth for these truths? We would like to do it, and we believe that they would receive people's reverence, if they were convincing to the end. But they are not, and their fate compels them to bolster the ranks of the "opposition". And this happens not only because they fail to understand that they are fighting against a myth imprinted in the DNA of the civilized world—growth. But they are also fighting against the form in which it has just started to gain credibility, thus entering the common universe of values—durable growth. In addition, they impose their truths using formulas that are imbibed in too much utopia to be able to persuade.

We believe that the excessive exploitation of the degrowth "aurora", Georgescu Roegen, has infused their views, premises, and analyses with too much pessimism. When you see that everything is burning and is transformed into "garbage" and nothing "gets in", you can only be sad.

And how can you promote such a work? Do you go about shocking, turning judgments upside down, announcing calamities and the end of the world? By fighting against claimed oxymorons and offering realist utopias in exchange? By assuming you can "stop" things that you know cannot be stopped? Who do you persuade that you only need an engine and not a car, be it even an infernal one? After centuries of waiting, do you hope to get an audience by proposing "prosperity without growth" and more jobs by "exiting the work society"? And then, on the same logic, you try to fit together two contradictory things: the wealthy should stop, while the poor should continue! If an economy firing on all cylinders, providing full satisfaction of needs, must stop, it does so according to the logic from the parable of the "rich man's stomach" that A. Smith irreproachable explained. There's no need for an aggressive radicalism, full of "stops" and "now", featuring "detoxes" and "exits"—from the material world, capitalism, sustainable development, etc.; with exits from all possible places! With the same parable, we can also explain very well the GDP-wealth relation, emphasized in the known Easterlin Paradox: beyond saturation, "your stomach cracks open" and wellbeing becomes pain. We would understand all this if we were referred to the *Theory of moral sentiments*. It is just that Smith is not solicited. From him and from the other classicists, missing in areas where they should be present, we would have received the necessary guidance to understand that it is not possible to improve living conditions without growth. Without it, relative inequalities would not be reduced and the number of jobs would not grow.

Anyway, when truths are deluged in an innebriaton of words, the chances of gaining recognition are significantly reduced. Such chances are also reduced by the wagons of utopia, phantasms and slogans in which degrowth drowns its assertions. A wealth tax on a global scale or a moratory on technical-scientific innovations (Latouche, 2009) are hilarious and are fed by the sorcery of a captivating world utopia. Starting from their wish to stop the "pleasure of living" and the preference for the present, the supporters of degrowth flagellate the market, competition, productivism and economism. They try to disguise some false and historically unproven assumptions to look like Popperian conjectures: that there is no real free time except beyond the market; that pluses of productivity have only been converted into increases of production and not into decreasing effort, etc. Any society that promoted growth, irrespective of its political colour, is considered guilty. It's true: capitalism looks more corrosive and greedier than the others. But even socialism can be "bourgeois". This is what Andre Gorz believes (Gorz & Latouche, 2009, p. 92). Thus, he and all the others vote for an "overhaul", for going "beyond", towards an eco-civilization that may be run according to an "ecological democracy". It's just as true that ecology should not have any political colour. But it does, and those who believe in an eco-democracy as well as in a non-ideologized state are just as utopian as when they ask for the abolition of the FED and the World Bank, the creation of a global currency, the abolition of interests, etc. The values promoted by the French Revolution needed a first consul, an "imperator", to be implemented. By saying a resounding NO to capitalist accumulation and to the "freedom to undertake and exploit", the supporters of degrowth activate the slogan of liberty, equality and fraternity. If generalized capitalism was to destroy the world, in what system and what way could the emblematic values of degrowth come to life? Would a generalized, subliminal, socialism be the right way? And if so, what kind of socialism? And in this show of the step towards "something else", who would be the "impartial spectator"? Without exception, the promoters of degrowth promote the asking of forgiveness! Let us first ask for forgiveness because we have chosen a wrong paradigm, so that afterwards, when we are "clean", we may embrace the right path. It seems that the "conscience" of the invisible hand can no longer play the role of the "impartial spectator". It was once entrusted to it and this is what we have come to! Anyway, the issue seems complicated. The impulse to think about other people's troubles and the negative effects of our own actions is here to stay. We have to share our fate, like brethren. And as Napoleon is no longer alive, the solution may consist of a global government in charge of a global Republic, with banks acting as compensation companies! This government will know what is moral and correct, and what makes everybody happy!

The supporters of degrowth do not lack self-critical sense. They wonder what they could be suspected of and, sincerely, they direct the attempts at answering that question to the area of obscurantism, reactionism or incorrigibly retrograde attitude. We do not believe that this is the cause of their failure to build a new world! And also, not because of a deficit in logic or temerity.

We believe, beyond everything else, that they will be suffocated by the sadness in which they dress their new "imaginary" and the radicalism they use to undermine the chance of a classical system that has proved its worth in surmounting the issues with which it is confronted.

We mean to say that the system conceived by the classicists, supported by the neoclassicists and rightly anticipated by the pre-classicists has turned growth into the magical word that expresses the will and the opportunity to solve the big and ingrate inequation of economic science: limited resources–unlimited needs. You don't have to study at Princeton or Cambridge to understand this. The fact that it is pitiable to send one's children hungry to sleep is enough for everybody to learn this lesson. If one is willing to accept that only through growth, the scarce things that one possesses will help one overcome the precarity of life, one reaches the same conclusion. People partake in the mythical nature of this conclusion. The myth possesses inflections and specificities, just as specific, diverse and complicated as the world that it addresses. It is still to be seen how the newly proposed paradigm, with the refusal of the material, is fit for this world; or, of at least two worlds. Contemplation and nice words cannot replace food, this is clear, wherever you are on Terra. Just as certain is also that a world used to eating well and sophisticatedly will not be convinced to switch to frugality and contemplation. On the other hand, holding a discourse about degrowth in Sierra Leone amounts to standing in front of a judge that has already signed the sentence. Here, material life has just begun and earned no brightness to give it up. The myth of growth hasn't been a historical constant. On the contrary, the world has dwelled longer in stationarity than in dynamic, not the kind of stationarity resembling the rest following a dash, but one considered in accord with wisdom and the balance of life. The change in substance and form came alongside the industrial revolutions, the invention of credit and accumulation, with the taste for investments and an ever better lifestyle. When people understood that they could live better through accumulation, investment, competition, private property and the free market, they didn't stop ever since. Growth became a religion and number one political pledge, both in capitalism and in communism. In the first form, it admitted all the intricacies determined by the presence of the Misesian "human action", but encouraged the entrepreneur. The latter form manifested itself by aligning minds, levelling, calculating and planning everything on each square foot. But mainly, by annihilating the enterprising spirit. This is the main reason why, in its case, growth failed.

The supporters of degrowth, without intending it, are about to act in the same manner. They do not drive the entrepreneur off the stage, but they also do not emphasize his role. Most often, the name with which he is addressed is "bourgeois"! We need not say whence this inspiration comes. In more friendly sequences, the institutional entrepreneur is given a pass. The forms it can take vary, from the engineer specialized in the recycling and reconversion of recoverable materials, to the mayor skilled in the "art of local rule" and even to the heavyweight entrepreneur - the state. If the free market is condemned for its short-sighted focus on the here and now, the eulogy of the clairvoyant spirit of the government, and its necessary authoritarianism, in compelling companies to devise social balance sheets and to present itself as the sources of wisdom seems self-evident. For important issues related to energy, distribution, climate, science, etc., the most suitable entrepreneur is still the government. Within and under the influence of such philosophy, the extravagant catastrophe-driven scholars of degrowth give advice. "They are close to us" not only regarding our choice, which we will have to make, of finding our matching country of the New North, sheltered from global warming, but also with respect to a country that knows how to make decisions! The societal perspective of the entrepreneur is not missing from the landscape of degrowth theory. He is driven to understand that common rights prevailing in front of individual rights amount to a transformation of himself. The intention is to make him, or first of all him, understand that profit is not everything. The common and general good would better define him!

This attempt, while essentially praiseworthy, is one of the great mistakes of the supporters of degrowth. They forget that the perspective of the enterprising investor is fundamentally individualist. If Latouche and Andre Gorz believe that the English investor cannot sleep out of concern for the Bengalese, then they shall have to live "pour toujours" the illusion of universal goodness. The investor is not indifferent with respect to the world, but his concerns have a different perspective. If he is interested in it, he does so because he wants to live in a world of peace, without pollution and, very importantly, a world with less risks for his investment. Thus, this is the second great mistake, which completes the first, of the supporters of degrowth. By announcing everywhere and at any time only ends of the road, imminent catastrophes, social, political and especially ecological turbulences, they draw a sinister picture. In such a context, risk and incertitude, Hicks' (1931) unwanted but present companions, remain obstacles for the investors. If those who outline such scenarios do not think about the effects on the propension for investments, the situation is dire. It's even more dire when those concerned are also economists.

Who is willing to invest when he thinks the world is ending or must cease to exist?!

What is the saving reasoning or philosophy that we might use to keep the entrepreneur on the scene? It is true, there are more than a handful of religious philosophies according to which the last days of one's life are more precious than the previous ones. Do we dare believe that the days will become very precious faced with the prospect of an empty and sombre future? That knowing our final hour we will make any mystery of effective demand disappear by buying and consuming everything that may come our way? And that against this background, the enterprising investors will profit by greatly exploiting this opportunity, as the very last of its kind? Who can guarantee that this will ever happen? Hicks most definitely won't. His world is full of risks, but acceptable risks, and it is not threatened by a merciless check-out. No, and he would tell us that under the permanently threatening sword of Damocles, the first to die is the enterprising spirit. And once that happens, the planet starts to "cool". If you are also inoculated with the notion that we would not wait long before we shall fight for the last slice of sun and the last square feet of earth not roasted by the threatening warmth, you can expect that nobody will spend their money on a business condemned to be frozen by a turbulent global warming or melted by a context whose emulative spirit falls down to zero.

If there is still a chance, with a high probability of fruition, it is that engineers mind their own business and do not read dreary stories about degrowth. It is only this way that they can use their energy and talent for saving technologies. It is only this way that they can build a super-sophisticated submarine boat like Triton, with which biologists can study the intimate life of the Great Coral Reef in order to come to its aid. Thus, protected from virulent threats, they conceive and build technologies to fight pollution and homes that are resilient when faced with earthquakes and typhoons!

Our opinion is that we must let them be and support them in their undertakings by encouraging them rather than sending shadows to haunt them. And we can do it by agreeing with the idea that growth is still necessary, very necessary, as well as desirable; it brings along the protection of environment, the avoidance and reduction of poverty, technical innovation, non-polluting investments and infrastructure, efficient human capital and, above all, trust in future. Should we remind you that one of the co-creators of the concept of ecological footprint, judging the example of China admits that growth is better than redistribution?

Against this background of mistrust, an enterprising perspective is pale or missing. If we were to advise the entrepreneur, we would say to keep a close eye on the calculations. Not the ones that prophesize degrowth, but the ones that balance the level of the ecological footprint against the chances of the economy to grow. If the maximum level of global warming is + 2°C, this is what they should think about in order to anticipate their chance at making a profit. This, we believe, is and will be the equilibrium equation of the future, an equation that makes the connection between the ecological footprint and the natural interest rate, which is itself an expression of profit estimated under the new conditions.

What about the rest? Well, we are left with the ascertainment that degrowth is an algebraic sum of inciting and fascinating questions and contradictory answers that are also full of ghosts. It's left for us to see that growth, even in its contested cloth of durability, is possible while avoiding disasters, that the emulating context proper for the enterprising spirit will follow the trajectory that links the ecological footprint to the natural rate of profit. And in addition, we should see that when reality is no longer served by the wagons of words that it uses, the theory under discussion ends in fallacies of logic, vocabulary, ideas and initiatives. They cannot give up reality; rather, it will have to get rid of the oxymorons that it constructed itself, for a better impression and less conviction. And with this occasion, they will realize that the discourse on sustainable development has been to an excessive extent confiscated by the problems of the environment. Because of this, it risks diluting too much the role of the main element of existence as a whole, of which nature is a part of, namely the economy, without which nothing can be built, repaired, recovered, etc.

References

Berr, E. (2006). *Le dévelopment en question(s)*. Presses Univ.

Besset, J.-P. (2005). *Comment ne pas être progressiste- sans devenir réactionnaire*. Fayard.

Blühdorn, I. (2015). *Post-ecologist politics: Social theory and the abdication of the ecologist paradigm*. Routledge.

Boulding, K. E. (1966). The economics of the coming spaceship earth. In H. Jarrett (Ed.), *Environmental quality in a growing economy* (pp. 3–14).

Cochet, Y. (2006). *Pétrole apocalypse*. Fayard.
D'Alisa, G., Demaria, F., & Kallis, G. (2014). *Degrowth: A vocabulary for a new era*. Routledge.
Dale, G. (2013). Critiques of growth in classical political economy: Mill's stationary state and a Marxian response. *New Political Economy, 18*(3), 431–457.
Daly, H. E. (1991). *Steady-state economics*. Island Press.
Daly, H. E. (2008). *Beyond growth: The economics of sustainable development*. Beacon Press.
Denis, H. (2016). *Histoire de la pensée économique*. PUF.
Diamond, J. (2014). *Collapse: How societies choose to fail or succeed*. Penguin Books.
Dobson, A. (2008). *Green political thought*. Routledge.
Fourier, C. (1846). *Théorie des quatre mouvements et des destinées générales* (Vol. 1). a la Libraire Sociétaire.
Georgescu-Roegen, N. (1971). *The entropy law and the economic process*. Harvard University Press.
Georgescu-Roegen, N. (1976). *Energy and economic myths*. Pergamon Press Inc.
Gerschenkron, A. (1962). Economic backwardness in historical perspective. *The political economy reader: Markets as institutions* (pp. 211–228).
Gevaert, P. (2006). *Alerte aux vivants et à ceux qui veulent le rester: pour une renaissance agraire*. Editions Ellebore.
Gorz, A. (1991). *Capitalisme, socialisme, écologie: Désorientations, orientations*. Galilée.
Heinberg, R. (2018). *The end of growth: Adapting to our new economic reality*. Seneca Lucius Annaeus.
Hicks, J. R. (1931). The theory of uncertainty and profit. *Economica, 32*, 170–189.
Illich, I. D. (2003). *La convivialité*. Ed. du Seuil.
Krugman, P. R. (2013). *End this depression now!* Norton.
Kumar, S. (2017). Spirituality and sustainable development: A paradigm shift. *Journal of Economic & Social Development, 13*(1), 123–134.
Kuznets, S. (2013). *Economic growth of nations: Total output and production structure*. Harvard University Press.
Kuznets, S. S. (1971). *Toward a theory of economic growth: With "reflections on the economic growth of modern nations."* Norton.
Landes, D. (1998). *The wealth and poverty of nations: Why some are so rich and some so poor*. Norton.
Latouche, S. (1998). *L' autre Afrique: Entre don et marché*. A. Michel.
Latouche, S. (2005). Vers la décroissance. Eco fascisme ou éco démocratie. *Le Monde Diplomatique, 52*(620), 1.
Latouche, S. (2009). *Farewell to growth*. Polity Press.

Lietaer, B. (2005). Des monnaies pour les communautés et les régions biogéographiques: Un outil décisif pour la redynamisation régionale au XXIe siècle. *Exclusion Et Liens Financiers: Monnaies Sociales, Rapport, 2006*, 73–95.

Mill, J. S. (1885). *Principles of political economy*. D. Appleton and Company.

Montel, O. (2017). La décroissance : Une utopie ? *Cahiers Français, 401*, 57–65.

Petit, O., & Villalba, B. (2010). Développement durable et territoires: quelques changements sans rupture. *Développement durable et territoires. Économie, géographie, politique, droit, sociologie, 1*(2).

Peyrefitte, A. (1976). *Quand la Chine s'éveillera...le monde tremblera: 2*. Fayard.

Randers, J. (2018). *2052: A global forecast for the next forty years*. Seneca Lucius Annaeus.

Raskin, P., Banuri, T., Gallopin, G., Gutman, P., Hammond, A., Kates, R., & Swart, R. (2002). *Great transition: The promise and lure of the times ahead* (Vol. 1). Stockholm Environmental Institute.

Rostow, W. W. (1990). *The stages of economic growth; a noncommunist manifesto*. Cambridge University Press.

Saigal, K. (2008). *Sustainable development: The spiritual dimension*. Kalpaz Publications.

Stengers, I. (2015). *In catastrophic times: Resisting the coming barbarism*. Open Humanities Press.

Tévoédjrè, A. (1982). *La pauvrete, richesse des peuples*. Editions economie et humanisme / Editions ouvrieres.

Vivien, F., Lepart, J., & Marty, P. (2013). *L'évaluation de la durabilité*. Editions Quæ.

Wade, N. (1975). Nicholas Georgescu-Roegen: Entropy the measure of economic man. *Science, 190*(4213), 447–450.

Wallerstein, I., Collins, R., Mann, M., Derluguian, G., & Calhoun, C. (2013). *Does capitalism have a future?* Oxford University Press.

Zaccai, E. (2002). *Le développement durable: Dynamique et constitution d'un projet*. Presses Interuniversitaires Europeenes—Peter Lang.

Zaccai, E. (2011). *25 ans de développement durable, et après?* Presses Universitaires de France.

CHAPTER 6

Nature – The Highlight of the Theory of Sustainability

The theory of sustainable development reserves a distinct and very important rubric for the environment. Apparently dry, this issue should belong exclusively to the positive level of science. What can be less ideological than air, water, oil, salt, etc.?! In spite of appearances, this is a locus of confrontations of economic, philosophical and doctrinal opinions. Starting from a common theoretical legacy that is succinct, partly incomplete but clear in dealing with the issue, the contemporaries have emphasized the labyrinthic and conflicting aspects; they have refined, updated, modelled and modernized; finding support in classical texts or, most often, disregarding them and building on their own. In doing so, some scholars have consolidated economic science, while others have ripped pieces of it or simply created other sciences.

6.1 The Dialectics of a Fundamental Relationship: MAN - NATURE - MACHINE

The complicated path of substitution and complementarity

There is no moment in the evolution of economic thought that failed to take into account the relation between man and nature. Up to the classicists, with few exceptions, nature is considered a whole; there was no

I. Pohoaţă et al., *The Sustainable Development Theory: A Critical Approach, Volume 2*, Palgrave Studies in Sustainability, Environment and Macroeconomics, https://doi.org/10.1007/978-3-030-61322-8_6

separation between resources and environment. Spearheaded by the physiocrats and identified preponderantly with agriculture, nature explained everything. Its sole serious competitor in the creation of wealth is labour. The philosophy of natural order also includes a strong mythical component. Earth comes from Nature. Hence the conclusion that creating clearly defined property rights for something that does not belong to you is a suspicious notion. Nature exists, it is present, it offers a playing field so that population, work and money may bring profit—this was the thinking of mercantilists. This means that, until the works of Smith, Nature was considered the mother originating from beyond this world; work was "earthly" and had to be eulogized. Real production factors make their appearance in the era of the classicists (Kiker, 1966). It is hard to know whether nature was included in wealth; rather, it served as a background and it was considered as given. We suspect this is what Montchrestien believed when he wrote that "The wellbeing of people consists, mainly, of wealth, and wealth is dependent on labour" (Montchrestien, apud Denis, 2016, p. 105). In any case, none of them, neither labour nor nature, can be replaced on the path towards wealth.

On the path of economic dynamics, the pre-classical mantra of the orientation towards nature is replaced with another mantra, that of labour. Labour confers meaning to the other factors as well. Ability, knowledge, skill, dexterity, health, etc. do not amount to much in the absence of labour—this is the common opinion of the classicists. Alongside them and through them and without stating it directly, the theoretical bases are laid for everything that would later feed the quarrel on substitution and complementarity; on the natural limits of resources and the self-development potentials of technical progress; on the tensioned relationship between man and machine induced by the former's replacement with the latter, etc.

In various grades and by discerning the unwanted possibility of stationarity, the classicists saw that the Earth is limited, as are natural resources. In relation to a growing population, this observation appeared to them as a major problem (Perman et al., 2003). The fixed land offer (Malthus), the diminishing returns (Ricardo), the compensating effect of knowledge (Mill) and the organic composition of capital (Marx) are the new concepts analysed. The pessimism induced by the limits of nature finds its solution either in the control of population or in free commerce. The market is identified as a means by which internal scarcity may be replaced by external abundance. The zero-sum game inspired by the objective theory of value is undermined by the theory of comparative costs. Any solution is good to secure the frictionless dynamics of the economy, by operating on every factor in part or in the space between factors. The differential rent brought about by the extension of cultures, but, especially, by artificially enhancing the quality of the soil are appealing to Ricardo and Malthus. This means that the Earth can get some exogenous help. Complex work may provide a substitute for simple work. Smith and Marx uncover the secrets of this process. Production is not only achieved by employing lots of unskilled people, but also with qualified people. The Smithian "knowledge fund" takes the stage! The always fresh originality which assures that a dead tradition is replaced by a live one announces, inspired by and through Mill's (2015) writing, the future "creative destruction". Hope comes from the "salt of the earth", namely, people's intelligence. Nature, education and culture empower work and make it more efficient. However, another thing is needed for the effect to appear. They need the "invention of all those machines by which labour is so much facilitated and abridged", Smith (1977, p. 24) clearly claims. Machines, instruments, in short, technical progress are called to *complement* and to assist *by easing* human effort; not so much as to replace it, but rather enhance it. From this perspective to the one where the machine and the tool are seen as a menace that threatens jobs, there is a long way to go (Kurz, 2010). Labour enhancement and its transformation into a veritable force of production were highly needed. The classicists are aware that the natural conditions of production, the area and quality of lands as well as the deposits of resources impose limits on production. All of them, without exception, are preoccupied by the effects that technical progress can bring in relation to labour and the "wage frontier". Marx's *Fragment on machines* and Ricardo's *On machinery* strike a different tone in this context. And they also do so from the perspective of the fact that, while fundamental, their texts are very little or not-at-all quoted by those that nourish the current discourse in the fields of strong or weak sustainability. Next, we will say a few things about Ricardo. As we consider Marx to be very interesting in this matter, we will allocate him a distinct paragraph.

In the chapter *On machinery*, that appears only in the 3rd edition of his *Principles*, Ricardo mentions almost everything we should know when we analyse, approvingly or disapprovingly, weak substitutability. Under the recognized influence of Smith, Ricardo strongly believes that the introduction of machines is a "general advantage", profitable both to the person who conceives of and uses them and to the worker. In the case of the former, reducing costs by way of growing productivity helps increase his profit; for the workers, "with the same wages in money" they can buy more goods, which are cheaper exactly because of the introduction of machines. It's just that, he admits, it's "often" likely that the introduction of machines would "deteriorate the condition of the labourer" (Ricardo, 2001, p. 284). This danger is only seen from one perspective: the firing of workers. And this may occur when "population will become redundant, compared with the funds which are to employ it" (Ricardo, 2001, p. 285). This is not a bias of the working class; political economy may scientifically elucidate the phenomenon, believes the quoted classicist. And Ricardo's Political Economy clarifies it clearly and definitively. The following quote is worthy of Descartes' spirit: "The statements which I have made will not, I hope, lead to the inference that machinery should not be encouraged. To elucidate the principle, I have been supposing, that improved machinery is *suddenly* discovered, and extensively used; but the truth is, that these discoveries are gradual, and rather operate in determining the employment of the capital which is saved and accumulated, than in diverting capital from its actual employment (…). Machinery and labour are in constant competition, and the former can frequently not be employed until labour rises" (Ricardo, 2001, pp. 289–290). In just a few words, but carefully weighted, Ricardo tells us a few fundamental things. The introduction of new machines is not done "without warning" and also, it does not produce "unexpected" results. A period of preparation is necessary, which includes saving money and investing in the novelty. When the new machines come, they "rather" complement than "withdraw" capital. The "destruction" is neither instantaneous nor on the whole line. Then, machines and technology, in general, do not exist on their own; they only operate "in competition" with people and never in their absence or by removing them. This means that they wait for "labour to rise"! In short, creative destruction is not as black and hostile regarding jobs, for objective reasons. Machines are not created for saints and they do not work by themselves. They operate with people who "rise". They grow up alongside the machines and come to meet the exigencies required by their creators. And the people who create them do it for a profit. By earning, they have the chance to save more and to invest more. This process does not prejudice anyone, because "These savings, it must be remembered are annual, and must soon create a fund, much greater than the gross revenue, originally lost by the discovery of the machine, when the demand for labour will be as great as before, and the situation of the people will be still further improved by the increased savings which the increased net revenue will still enable them to make" (Ricardo, 2001, p. 290). An additional benefit, in Ricardo's view, would also come from the chance of a more efficient external commerce. Competitive advantages are gained through goods supplied at low costs by way of technical progress. The warning that a capitalist would cross the borders with his capital if he were hindered in promoting new technology for fear of unemployment remains a concern. This was the case in his time and it still is today.

In short, the relation technical capital–labour is seen as a causal relation with a circular flow. The level of wages, the level of occupation and the interest for profit define a mechanism in whose operation everybody is interested. Labour substitution is not even remotely an issue; there is no "withdrawal", but rather addition, multiplication, if not within the same boundaries, then definitely within others. But technical progress does not multiply by itself. This process has to be animated by people. Change might be an inconvenience, but one that pales in comparison with the achievement of "general good, accompanied only with that portion of inconvenience which in most cases attends the removal of capital and labour from one employment to another" (Ricardo, 2001, p. 282). Period!

What does this self-propelling and resilient mechanism lack? Apparently, it lacks nature. Or, at least, this is the common perspective. In fact, it doesn't. It is present in the range of circumstances that lead to final results, both in agriculture and industry. And in this exercise, what matters is not the limits of nature, but rather the efficiency with which the three main factors are used—the productive triad. Geography matters too, but not primarily. Smith (1977, p. 12) postulates that wealth is "in every nation (…) regulated by two different circumstances; first, by the skill, dexterity and judgement with which its labour is generally applied; and, secondly, by the proportion between the number of those who are employed in useful labour, and that of those who are not so employed. Whatever be the soil, climate or extent of territory of any particular nation, the abundance or scantiness of its annual supply must, in that particular situation, depend upon those two circumstances". This "whatever" does not mean *lacking*, but rather *under the conditions given by…* Whether nature costlessly participates in general wellbeing or not, that depends on the place where labour is expended. In his *Principles*, Ricardo claims something very important. He states that "natural elements" like air, water, wind, air pressure, vapour elasticity, etc. are *inexhaustible* and, which is worthy of noting, nature works for us free of charge. The use of such elements does not involve any expense; they *do not have a price*! On the other hand, natural elements do not produce anything in the factory, but rather man does everything, Smith believes. Taking another stance, Ricardo believes that nature contributes here too. The element of *scarcity* is not absent from this analysis! Nature's "labour cost no expense"—Smith claims, without thinking of any limit. It bears no costs, but it is paid when its gifts become scarce (Ricardo, 2001). Gratuity only operates when gifts are in excess! Rent, in its turn, gives rise to different opinions. It is the work of nature both for Smith and Malthus. For Ricardo, rent is not "the great gift of nature", "the force that determines grass to grow and grain to ripen". Anyway, for all of them, without nature labour does not create any revenue. *The notion of substituting nature in any way seems absurd.* If production, mainly agrarian and extractive, encounters any limit, the solution comes from the free market, be it internal or external. But it too comes to complement, not to replace. Namely, they have in view the dynamics effects of markets in the current terms of the discourse on the matter. Technical progress contributes too. It possesses the greatest capacity to ease scarcity. This is because "resources can only be defined in terms of known technology", as correctly pointed out by Harold Barnett and Chandler Morse (1963, p. 7). None of the classicists proved that they knew the power of technical progress in discovering new resources or in finding the most efficient ways of exploiting the known ones. Hence some of their pessimism. Malthus was the saddest, as he also saw the absolute limitations of resources. Ricardo was just sad, but not hopeless, as he saw commerce as a saving compensatory force.

Mill's "naturalism" strikes a very particular tone. His view is that man is a part of nature and in no way a divine creation. Therefore, the relation between man and nature does not go through God. It is in the power of men to solve the problems. His stationarity is not catastrophic. A stay does not mean a definitive stop or the impossibility of reaching a farther target. Noteworthy is also the idea of the implicit distinction within nature. Earth with its limited areas and various fertility rates is perceived as a resource. This includes mines too. The other "elements of nature" define the environment! This is what Mill is talking about when he prescribes recipes for arrangement and embellishment!

What is there left to be added? Today's sustainable development may find inspiration in the model of Ricardian intensive agriculture using natural fertilizers and crop rotation. The temporal order of primacy of the two factors—capital and labour, which is still subject of debate—has found its definitive clarification in the works of the classicists. Smith is reliable when he claims that technical capital—machines—are the results of labour. The fact that by using perfected machines, labour gains additional capabilities to form a larger and more valuable capital does not change the essence of the problem. The path always leads from labour to capital, never the other way around. If there is anything to discuss, it is the way in which the two forms of capital ***coexist*** and how labour—human capital—is condemned to "complementarity with technology" (Fuente et al., 2003).

What do we have so far? We see that the classical route of the productive triad is explained in terms of cohabitation and complementarity. In the works of pre-classicists and classicists, with the exception of Malthus, there are no absolute limits to growth. When an impediment is reached, machines and the market come to help. The price formed on the free market may be a valuable indicator: for the entrepreneur in using his capital, in a competitive and highly efficient manner, on domestic and foreign markets; for the worker in finding out how strained his wage relation is in relation to his employer and to the welfare allowed by the promotion of new technologies. In no way does price offer information about the relative dimension or the quality of resources and it also does not suggest the need for any substitution. Technical progress is also a solution to scarcity and it is a way to gain competitiveness. If we fail to employ it, we lose in relation to others that did. The Malthusian scarecrow of massive unemployment lacks rational support. In addition, technical progress is necessary to ease and streamline the human efforts of both entrepreneurs and workers. Its refusal has no logical justifications, and it is Political Economy, rather than any other science, that must explain this. It has to explain one starts from labour to amass capital and that without nature, labour is consumed "in a void". We also keep in mind that in the pre-classical and classical scheme of development, nature "receives" and "gives". From this point of view also, the necessary harmony calls for equilibrium. Furthermore, the naturalism of pre-classicists and classicists is present. From its perspective, the relation of man with nature is not ethically and morally castrated. The classicists were no standard naturalists or ecologists. But they were also not absent from any of these two rubrics.

What is grafted on this scheme by the neoclassicists? It is grafted and in quite a consistent manner. Their aridity on the level of positive science, where they excelled in modelling and dry marginalist calculations, has often created a false image. It is only by letting go of this image that we see how neoclassicism offers undoubtful evidence that it is not estranged from the environment and the issue of material and human resources. Menger, Marshall, Pigou, Jevons, etc. have laid foundations in these areas too. In the following sentence, Menger shows how important the human resource is in the process of creating value: “Man himself is the beginning and the end of every economy” (Menger et al., 2011, p. 100). In just one sentence, the same author talks about the role of the progress of human knowledge in promoting general economic progress: “Nothing is more certain than that, the degree of economic progress of making will still, in future epochs, be commensurate with the degree of human knowledge” (Menger, apud Burns, 2018). The human needs-scarcity relation is commented upon by Johnson to outline the contribution of Menger and the scholar inspired by him, Abraham Maslow, in constructing the famous pyramid of human needs, a derivative of goods’ scarcity (Johnson, 2017). Gareth Dale also considers Menger’s contribution to the analysis of the concept of scarcity as a valuable one (Dale, (2017). The attention awarded to scarcity and the way in which a noneconomic good abundantly available (air, water, natural landscape, etc.) can become an economic good “by deliberately producing scarcity” is also a Mengerian idea analysed by Foster and Clark (2009). Cordato retains the pioneering role of Menger in establishing the importance of property rights in solving environmental conflicts (Cordato, 2004).

The quotations from Marshall and Pigou as unwilling founders of the future *Environmental Economics* are constantly present in the analyses of environmental issues. Marshall is interested in the study of work and progress, but also in that of poverty. From this perspective, he serves as inspiration for A. Sen (1988). Marshall’s progress “must be slow relatively to man’s growing command over technique and the forces of nature” (Marshall, 2013, p. 207). This Ricardo redivivus is noteworthy! “Machines grow” slowly, together with people, in logical cohabitation rather than disjointedly. The status of nature’s “free gifts” is not indifferent to him! They were “partly impoverished and partly enriched by the work of many generations of men” (Marshall, 2013, p. 122). At the public conference *Pressure of population on subsistence means*, London 1885, he claimed that three sets of physical needs support a wealthy society: raw materials, industry, but also clean air and water plus areas for recreation (Caldari, 2004). It is an opportunity to also claim that demographic pressure is an enemy of fresh air and open space. He has no reservation in proposing a "tax on clean air" necessary to create green spaces in the middle of crowded industrial districts. What else is there to say? We should say what everybody knows, namely that Marshall's name is linked to positive externalities and the chances for increased profits of a company that is operating in an « industrial district», thanks to the "general atmosphere" of the future externalities. It is also known that Pigou develops his idea by proposing the taxation of negative externalities as a means to *make up* for the effects on the environment. This idea will be made famous by the well-known principle "the polluter pays". Famous will also be the controversy, also originating in *The Economics of Welfare*, between net social marginal product and net private product which, in the end, will find its synthesis in Coase's social cost. In essence, it was also an attempt to save the market. Its deficiencies in creating externalities, doubled by the need for state intervention to correct them, find the solution in a negotiation between the polluter and the polluted. Coase invites them, in a democratic manner, to freely negotiate, with only one prerequisite: clearly defined property rights. This notion will dissatisfy all later ecological economists when they would deny any property right over the biosphere.

Jevons is another scholar on whose example one can prove that not all the neoclassicists deserted the notion of the specific character of natural resources, operating only with perfectly replaceable defying homogeneities. He has not entered the stage of Economics by publishing a *Problem of natural resources*, but with the famous *The Coal Question* (Jevons, 1866). This means that he talked about something specific, coal, a resource whose ending he was predicting. He thought about this ending in physical terms, even though he also crafted analyses in terms of price. The direct negative rebound effect is an example: the lowering of the price of products obtained by using the resource after growing the efficiency in exploiting it increases its consumption (Jevons, 1866). Dale (2017) appreciates that even the current issue related to "peak oil" may be analysed from Jevons's perspective. It is the line of thought started by Jevons that would later inspire Hotelling (1931). By way of his scarcity rent (the Hotelling rent), he would send this issue towards the market and, with its help, he would "link" generations by the specific mode of calculation he proposed: the market value of an exhaustible resource should be linked to the stock of resource not yet exploitable. Or in other terms, the higher the scarcity of a resource, the higher its price, regardless of the real exploitation price. With all the many criticisms addressed to him (neglecting the effects of technical progress being the most severely taxed), the Jevons-Hotelling idea remains present in all the analyses of *Environmental Economics*.

The neoclassical founders were neither more ecologically inclined, nor more naturalists than the classicists. It's true, they lacked the Malthusian fervour in talking about the limits of resources, but they were not absent in this regard. The perpetual struggle for land and food were for Malthus indicative of the limits of resources (Malthus, 1983). Jevons found them just visible, without needing sophisticated studies. The Malthusian solution indicates an adaptability for survival à la Darwin (Galor & Moav, 2002). Jevons and the other neoclassicists pin their hopes on the market; they think of the purity of air and water, as well as the beauty of urban areas, just as Mill thought about the rural ones. The clouds of finitude and the limits of the earth preoccupied them without troubling them. Were they more preoccupied with the efficiency of allocation than with the limits of resources? It is difficult to say! However, if they did, it is not the analysed scholars who are to blame. Maybe their followers are. In any case, it is not at the level of the founding neoclassicists that one should dramatically ask the question: Why or what should one substitute? The scarcity of resources was a problem, but it did not cause anxieties! By the theory of comparative costs, the classicists have also proved awareness of the unequal distribution of resources on the globe. And they saw the market and free trade as convenient solutions. The neo-neoclassical versions of the Ricardian theory do not leave this paradigm. However, for ecological economists, the unfair manner in which the original geography has distributed resources is a problem. It does not help that there are, in general, resources. Their "territorial" setting is very important (Zaccai, 2002). The question is not how much population, in Malthusian terms, the Earth is able to feed. The real question is how resources are territorially distributed in relation to the distribution of population. The translation of the analysis of resources from physical to price-related terms has not dealt with the excesses. Even the standard judgement that the exhaustion of a resource drives up its price thus making it worthwhile to exploit, in terms of efficiency, another resource does not belong to the generation of the founders of neoclassicism. Also missing is the pretence that a higher price tag allows for the discovery of more resources!

And so, where does the theoretical-doctrinal radicalism in this matter come from and against whom does it manifest? Where does the classification of economists into economists and ecological economists come from? Where do the disparities and the great division between strong and weak sustainability come from?; the intrinsic and unique value of a finite nature that cannot be substituted vs total substitution between natural factors with the ones created by humans?; complementarity vs substitutability?; the erosion of natural capital vs the compensation with other types of capital?; the risk of exhausting resources vs the confidence in self-development supported by the market?; readjustment vs the socialist transformation of capitalism?; Roegen, Daly, Cobb, Constanza, Martinez, Passet, etc. vs Solow, Stiglitz, etc.?; finite nature vs plentiful nature; wealth achieved through resource preservation vs wealth created by exploiting them?; the Economics of environmental resources vs Ecological economics, etc.?

In any case, we have sufficient reasons to think that the founding neoclassicists cannot be the target of such interrogations. What about their descendants? Who are they and who identifies them as such? Who decides today the classification of economists into the followers of classicists, on the one hand, and the followers of the neoclassicists, on the other? Nobody does, and it would not be proper to divide them according to this criterion. And with respect to such issues as the ones above, especially environmental issues, rational judgements must prevail over doctrinal affiliations. A communist who loves the environment may be just as likable as a liberal in love with nature. It's also true that this does not apply with respect to the usable instruments. But do we reproach anything to Solow or Stiglitz because they represent today's neoclassicism or because they take up too much space on the stage and they represent a mainstream that does not see anything beyond models? No, we don't! We will see that these two scholars, and not only them, defend their ideas and arguments without assuming any clearly marked doctrinal position. They are interested in common-sense arguments originating from the logic of things rather than from any gloomy doctrine. They are credible not because they are neo-neoclassicists, but because they have vision and tempting reasonings. And, in addition, they also enjoy clean nature.

6.2 Substitution in the Vision of the Theorists

We have titled this paragraph as such with the thought that there exists another kind of substitution. One that belongs to engineers and innovative creators who invent without having read Solow or Dasgupta. And they do it by competing with one another and with a single obsessive consideration in mind: to ease the work of man and to make life beautiful. They do not fight one another. With theoreticians it's somewhat more complicated. Even when they settle on common objectives, they do not use their energy pursuing the same issues. On the issue of HUMAN-nature relationship, what is it that parts them so much? The principled answer should be: two philosophies rather than two doctrines, with different supports and predictions. One is restrictive, considering that human actions take place within a finite space; according to the optimistic philosophy, they have limits; however, it pins its hopes on human ingenuity to overcome them. From the concise inventory of ideas we have put together, we can see that neither the founding classicists nor the founding neoclassicists can be identified, with clear certainty, as the source of the theoretical split that happened in the 1970s. Who of the classical founders can be quoted by Georgescu-Roegen, Herman Daly, Kozo Mayumi, Michael Jacobs, Peter Soderbaum, Tim Jackson, Martinez-Alier, Giorgios Kallis and the others to consolidate the line of hard sustainability? A few or none at all, except Roegen who explores the domain of Smith-Ricardo-Malthus and Marx; for the others, the category of classical references is scarce. Only Malthus finds his place, either directly or by quoting from other sources, in the works of Daly or Martinez-Alier. The other line, of soft sustainability, is just as frugal when it comes to classical references. Malthus is quoted by Stiglitz and Malthus plus Ricardo are mentioned by Chandler Morse and Harold Barnett. Solow, Hartwick, Dasgupta, Geoffrey Heal, Wolfgang Buchholz, Tapan Mitra, David Pearce, Giles and Atkinson build on their own, without classical references. However, the neoclassicists Menger, Marshall, Pigou and Jevons enjoy more attention.

In total, the classical paradigm of scarcity may be appropriated by the supporters of hard sustainability, but it does not provide a ground for radical-fatalist attitudes. We have seen that even the most pessimistic, Malthus and Ricardo, contemplated such solutions as were befitting their own terms; however, they have not suggested a universal lament. The stationarist Mill does not do so, either. The masochism of an analysis according to which by staying still or by slowing the rhythm of growth, we would consume the planet anyway, does not come from him. The traditional neoclassical paradigm is also far away from any radical attitude. Jevons' prediction was confined to history. Marshallian and Pigovian externalities are serious and welcomed warnings for building a civic attitude of getting involved in ensuring environmental health. The correspondence between needs and the scarcity of resources also remains a lesson that Menger did not shout about.

Thus, where are the mentors? We believe that neither Peroux nor Schumpeter, with their distinctions between growth and development, can feed a discourse with fatalistic accents. The self-propelling tendencies of a dynamics supported by the free market à la Smith via Rostow (1990) may indeed emulate self-development of the Solow paradigm. But beyond scholarly sources, we believe that factual realities of the years when the theoretical discourse became radical had a determining role. The more and more prominent signals related to the crisis of the renewable resources, pollution, global warming and in general the environmental footprint, the unemployment of graduates and the persistent poverty in some areas of the globe, etc. have created the context and the pretext. The awareness of this fact didn't wear the same garment. Kuznets treated it academically. Georgescu-Roegen set himself as the head of the hard line: economy is wicked physics, and unless we quell our pleasure of living, the planet splits off! Solow places himself on another route: this is true, but there are possible solutions for the future generations not to feel sorry that they were born from irresponsible parents lacking universalist worries. He and the scholars who share his opinions believe that substitution may still work. At the crossroads of these two great turnings, another one with an exceptional development comes to complement the dispute. It's not just geography and resources that matter, as J. Diamond believes! Institutions are even more important, Acemoglu and Johnson claim.

Who are we to believe? Who is more sustainable in their arguments?

6.2.1 *Strong Sustainabilists: Natural Capital is Indestructible!*

The turn towards the physiocrats seems to be the first urge of the scholars who give the signal of the inevitability of Ecological Economics. Physiocracy has put nature at the forefront, which is the only creator of new values with divine support, in relation to commerce and industry. In this sense, it is reliable to invoke it as a background for ecological development. But that's all! The physiocratic natural order is not gloomy! It is divine, and its mechanics is the laissez-faire and free market, with the pursuit of personal interest. However, things are not the same on the other side: the market is guilty of everything; it is the boxing sack of all ecologists. However, this is the philosophy of natural order on which the bases of the new hard sustainability paradigm are set. Under its protection and in the name of "the common patrimony of humankind", natural capital is defined, which is a key concept for strong substitution. Until Herman Daly would define it, the field was prepared by Vladimir Vernadsky (2002) with his notion of the biosphere as an indestructible and indivisible whole! In his turn, Georgescu-Roegen crushes the market of Marginalism, which was at its peak, and places under a dramatic question mark the possibility of physical self-development. If the piece of coal burned indefinitely and the conversion were 100%, we would live without a care in the world. However, the second law of thermodynamics tells him, and then in his turn he tells us, that we live in a finite, clearly marked world, and that the admittedly human pleasure of living is the original message about economic insufficiency. Herman Daly receives his "baptism" in such waters. In tune with his mentor, he writes that "low entropy is the common denominator of all useful things and is scarce in an absolute sense" (Daly, 1991, p. 41). Natural capital receives its significance from the same theoretical framework. It is "a stock that yields a flow of valuable goods or services into the future" (Costanza & Daly, 1992, p. 38). His embroidery on the distinction between "stock", "service" and "flow" is meant to tell a single story: capital is only a transitory phase of natural inputs, a modality for their transformation into utility, and nothing more. This means that the process does not have anything to do with productivity. It does so for Solow, but not for him or his other colleagues. And the process of transformation can only be a corrosive one, diminishing resources and reducing non-human species. This translates to costs and minuses transferred to future generations (Daly, 1987).

By meddling with the common heritage of humanity, the ethical dimension of the issue is strongly emphasized. Not only the harmony with nature, but also the correctness of the manner in which each generation consumes from the same limited whole appears as a first-order ethical matter. Capital and resources ("flow" in Daly's model) cannot be substituted. Resources offer only the material base for the existence of capital, while the latter has no productive contribution beside the generation of services for individuals, as the elements that make it up need maintenance to keep producing effects and are annulled in the long run because of physical decay. This is a Malthusian scenario which is also employed by Ecological Economics. It is a scenario in which the entire base of resources available are known from the very start, the level of technological advancement is given, and society can last as long as the ratio between production and resources is maintained at a low level. The law of entropy suffocates growth, with the specification that Daly finds a trace of hope in the chance for a qualitative development against the background of a quantitative stagnation of growth. On the other hand, the same Daly, as well as the other ecological economists, considers that price stands no chance of solving the issue of absolute scarcity. However, it does solve relative scarcity, suggesting a possible substitution between technical factors; but it has no power with respect to absolute scarcity. And the problem of Daly and Ecological Economics is absolute scarcity. Here, in this area, goods become the subject of positive elasticities, just like Veblen's luxury goods, generating validating mechanisms for Jevons' rebound effect. Rising prices do not lead to a fall in demand, quite the contrary! It is true, this happens under conditions of increased efficiency, a situation that is foreign to the analyses of ecological economists. Foreign and rejected is also the hypothesis of recycling as a chance of reducing absolute scarcity. In the same manner, they reject every hypothesis and effort to think about a substitute for the biosphere (Daly, 1973; Wackernagel & Rees, 1996). Hence, the denial of the use of production functions with natural resources, in flagrant contradiction with the law of entropy, and the refusal of every function that would accept the hypothesis of substituting natural capital, which is rare and limited in physical terms. This is blasphemy, as all ecologist economists believe.

6.2.2 *The Solow Approach—self-Development Through Ability À La Brundtland*

The Solow model of development is considered a standard neoclassical reference. Serving as the support of several theoretical developments, both pro and con, it brings together, but is also completed and rounded by Hotelling, Hartwick, Stiglitz, Nordhaus and Dasgupta.

In spite of appearances, it does not suffer from the ailment of dry mathematics. The daring attempt to find substitutes for natural capital within man-made capital, also provides a trip through a causal network of consumption, savings, investment, natural resources, interest rates, the level of population, time and economic efficiency. Ethical accents also seriously emphasize the analysis of the relation between generations.

The philosophy of the Solow model cannot be framed into the model of the standard notion of sustainability; it is a "buzzword", he believes. It is a notion that belongs rather to "distributive equity" than to issues of economic efficiency and compels one to not be "morally obligated to do something that is not feasible" (Solow, 2000. p. 180). It is neither sensible nor necessary to deliver the world to the next generations as you have found it, because we have no clue about their costs, their preferences, and the technology they will use. In other words, "it's not our job" to deal with something that we do not own. In terms of this hypothesis, Solow the "antihuman" and the "enemy" of all ecologists come to the well-known and controversial conclusion that "[t]he current generation does not especially owe to its successors a share of this or that particular resource. If it owes anything, it owes generalized productive capacity or, even more generally, access to a certain standard of living or level of consumption" (Solow, 1986, p. 142).

The refusal of the notion of preserving resources in physical terms for our descendants does not amount to an attitude of indifference. The special mode of raising the issue regarding the contents of the relation between generations infuriates the scholars who trace their roots back to Roegen, but it does not go against a certain reasonableness or the Brundtland spirit. And the said Report talks about "ability". This is Solow's "capacity of generalized production"; we convey the "fishing rod" to the next generation, but not the already caught fish.

To make his opinion a source of inspiration for contemporaneous decisions, Solow moves things on a field proper for a typical marginalist analysis. He translates the issue of resources from physical to value terms. The price, as well as the interest, become key working concepts. Building on the logical structure offered by Hotelling, he links the value of a deposit of resources to the interest rate. The main issue then becomes one of updating, rather than one of resource transfer. Petty's uninvited spirit is present here: the rent value set at the level of the current interest rate leads us to the price of land! Building on this logic, Hotelling believes that, in a state of equilibrium, the value of a deposit of resources must go up by a rate equivalent to the interest rate. Not far away from Petty, but also not from the classicists, Solow considers price and the demand–supply mechanism the main determinants that decide the opportunity of exploitation of a resource or the lack thereof. When the market price exhausts a resource, the same mechanism makes another resource attractive. W. Nordhaus extends the judgement and gives credibility to the hypothesis of substituting an exhaustible resource with an inexhaustible one. A judgement à la Böhm-Bawerk is welcomed: if "market interest rates (are) to exceed the social rate of time preference (...) then the market will tend to consume exhaustible resources too fast" (Solow, 1974b, p. 12). Hence the implicit conclusion that choosing a social discount rate is highly significant in shaping the policy of distributing resources among generations. The question of who wields the directing baguette of such a mechanism is also important. "A market-guided system" (Solow, 1974b, p. 12) is Solow's concession; a little bit of governmental assistance should do no harm! Choosing the type of macro-management—laissez-faire or state intervention—gets more complicated when the discussion includes the type of property over resources. It is well-known that the time preference bears the mark of rationality at the individual level. Does the government make correct intertemporal decisions supported by state preferences?! Might possible conspiracies against the public interest resuscitate A. Smith's warning (Solow, 1974b, p. 14)?

The preference for time is also important from another perspective, one concerning the dynamics of consumption. Solow and the other supporters of soft sustainability are concerned with the future of consumption. And good chances do not arise out of present abstinence. Quite the contrary, "a society that invests in reproducible capital (...) will enjoy a consumption stream constant in time" (Solow, 1986, p. 141). A reference to Hartwick may consolidate this idea. Investing a part of the profit extracted from the exploitation of natural resources in reproductible technical and human capital—this is the way by which a society may maintain a constant flow of consumption (Hartwick, 1977, 1978). The objective—constant consumption—involves a mandatory rule: saving-investing. The rents generated from the exploitation of exhaustible resources must be transformed into reproducible capital by applying this rule. Against such a background, Solow's conclusion is obvious: "accumulation of reproducible capital exactly offsets the inevitable and efficient decline in the flow of resource inputs" (Solow, 1986, p. 145). We do not need complicated policies! If the rents from resources are always invested in reproducible capital, then we shall happily ascertain that "some appropriately defined stock is being maintained intact, and that consumption can be regarded as the 'interest' on that stock" (Solow, 1986, p. 146).

Until what point can we go on like this? Up to infinity, Solow suggests, unless we tie ourselves to the mast like Ulysses to ram into unhealthy temptations and we fail to leave aside the naiveties of the Club of Rome! "What would Solow say?" Hartwick (2008) was also drawn to wonder! And this is the possible and assumed answer: "[i]f it is very easy to substitute other factors for natural resources, then there is, in principle, no 'problem'. The world can, in effect, get along without natural resources" (Hartwick, 2008, p. 91). A world without problems! All the natural resources can be replaced! There is no dependence on planet Earth! This is a construal that is hard to believe, almost utopian! In contrast, reliable and belonging to the area of hard sustainability is the idea of the same Hartwick, this time expressed in a realist manner, that "the correct measure of NNP, a measure that incorporates the current loss in value of natural resource stocks due to use for exhaustible resources and, roughly speaking, over-use for renewable and environmental resources" (Hartwick, 1990, p. 292).

Connecting the issue of natural resources consumption to the paradigm of the relation between generations, Solow ascribes it a special ethic and acknowledges a double sense to it. It's not only important what the old generation does for the new generation; the reciprocal is also interesting, he warns! From this point of view, time can be "a useful trick, reflecting the 'physical' fact that there is no way the past can be compensated by the future after the saving has taken place and the productivity of capital goods exploited" (Solow, 1974a, p. 34). In other terms: unless you live your moment, nobody gives you any other chance to live it; thus, live, consume from what nature offered you, this seems to be its command! Saving or not saving on behalf of a generation that follows doesn't seem to be a rational solution. Nor does the maintenance of an intact stock of environmental resources given by nature. And this is because, certainly, "technical progress would favour the future over the present" (Solow, 1974a, p. 33). The max–min criterion is also subject to the same logic, except for the case in which the initial stock is precarious and poverty may be perpetuated by following the rule of primacy of the present good. For the rest of it, " earlier generations are entitled to draw down the pool (optimally, of course!) so long as they add (optimally, of course!) to the stock of reproducible capital" (Solow, 1974a, p. 41). In other terms, it's reasonable that every generation takes care of itself. The equity of the present moment is just as important, Solow seems to claim. From this point of view, the control of population growth appears to him in terms of a good policy to the benefit of sustainability, especially for the third world.

The population level, but especially its dynamics, is a variable that interests Stiglitz too. In his attempt to tame Solow's model, he engages into the fight against the limits of natural resources with three compensatory forces: technological change, the substitution of natural factors with artificial factors, and returns to scale! (Stiglitz, 1974). Starting from here, Stiglitz postulated in five main sentences what needs to be done to maintain a constant flow of consumption and to start and maintain a self-development mechanism in which outputs do not exceed inputs; the inputs of natural resources should be covered by the outputs of human and technical capital. Of special significance is a sentence which includes the population variable: "[i]f the rate of population growth is positive, necessary and sufficient condition for sustaining a constant level of consumption per capita is that the ratio of the rate of technical change, to the rate of population growth must be greater than or equal to the share of natural resource" (Stiglitz, 1974, p. 128).

What else complements the "Solow atmosphere"? Is it Hicks' intuition of self-dynamics? Yes, united together, the three names Hicks-Hartwick-Solow define the rule of a perfect, very light sustainability between the different types of capital. This line of thought also includes Chichilnisky (1996, 1997). His "mixed" criterion attempts an appeasement between the "dictatorship" of the present and the "dictatorship" of the future. The actualization of the utility level of each generation appears to provide a solution to the intricate dilemma efficiency-impartiality. Let us not forget about Dasgupta and his and Buchholz's appreciations of Hartwick and Solow (Buchholz et al., 2005). In accord with Nordhaus and emphasizing the special role of non-renewable resources, he suggests together with Heal that "the easier it is to substitute, and the more important is the reproducible input, the more one wants to substitute the reproducible resource for the exhaustible one" (Dasgupta & Heal, 1974, p. 12).

6.2.3 *Can Sustainability Be Nourished from the Solow-Daly Querella?*

A modern economy does not raise this issue only in physical or monetary terms, both are taken into account! The transposition of judgements from physical terms into value terms seems to be, from afar, the main thrust of soft sustainability against hard sustainability. And this happens against the background of the "heroic" hypothesis of the presupposed homogeneity of resources! Between these two front lines, local "ambushes" define the general conflict. Biosphere and the physical environment in general are fixed data for Roegen's followers. Consumption leaves an empty space behind it. For the scholars on Solow's side, nothing is fixed as long as technical progress, by infusing everything, extends and develops, if not quantitatively, surely qualitatively. What is taken away may be replaced by human creation. Mutations are present in the structure: when the share of natural capital decreases, the share of created capital increases! On the edge, even scholars of weak sustainability acknowledge that not everything is sustainable. A welcome addition comes here from Toman M.A. He suggests a classification and an a-priori separation of things that can from things that cannot be substituted, trying to reconcile the neoclassical criteria of the optimum with the eco criteria of conservation of the "critical natural capital" (Toman, 1995). Anyway, future generations "have a choice" between receiving more productive capacities or more untouched nature! Solow thinks that this judgement has pros and cons and that future generations will know how to take care of themselves. His mathematical model has convinced him of that. "Transitivity is mathematically convenient but it is not necessarily a property of the real world", J. Gowdy replies (Ayres et al., 1998, p. 6). He also believes that the Jevons paradox checks and thus, by changing prices and efficiencies, we do not solve the issues of a physically finite world (Mayumi et al., 1998).

From this analysis, we have seen that Solow, Stiglitz or Nordhaus do not deny nature, but they try to pin it between different coordinates. They try to link generations just as Bawerk did, by means of interest! Instead of delivering the planet physically intact, we should rather update its value with the current interest rate! In addition, as a rule of prudence, we invest in transmissible capital, a part of what we get from exploiting nature! Costanza and Daly (1992) see prudence differently, not within min–max models, but rather using "maximum sustainable earning rates". Daly coats the concept and sees forethought in the compliance with a few rules: the exploitation rate equivalent to the regeneration rate; the rate of waste emission equivalent with the capacity of recycling and absorption; the rhythm of exploitation of finite resources equivalent with the rhythm of their possible substitution with renewable resources (Daly, 1990). We can see here a concession in relation to Roegen, who did not admit that recycling had any virtue at all. Admittedly, Hartwick's rule spurs prudence too. What is the permanent investment of the rent from the exploitation of natural capital if not a good practice that defines moderation and prudence? On the other hand, the process of updating by using the interest rate instrument does not seem an enterprise with aims for prudence, but rather an exercise in support of the market's immediate purposes. One does not update when one wants to conserve, but rather when one wants to sell! Selling does not necessarily mean taking the object outside the national borders (depending on the resource), but it certainly means its delivery for exploitation. However, an interest rate may be proper in the fight against inflation, but not in support of a strategy dealing with resources. Resources constitute the object of strategic decisions not because they represent values, but because, first of all, they are exhaustible physical reserves. This is why the intention to exploit a resource belongs to public power. The residual rights of control and the hybrid institutional arrangement state-market in affecting and exploiting resources have a word to say here. In the liberal eyes of the neo-neoclassicists, the market seems to be the most adequate institutional arrangement for the guaranteed allocative efficiency. To guide what is obtained from the exploitation of resources towards productive capacities by the public force while sending under the tutelage of the free market the way of attracting resources, such a scheme would be subjected to the charge of hypocrisy from those on the other side of the barricade. Is that so though? We think not, based on the facts that on Solow's market there is a place for the links between generations or the faith in the durability of an economy built not just on any ethics, but rather on one that is open to recognized human values. There is no hypocrisy here! Solow's model excludes neither the present nor the future man. It's just that for the future man, he has another solution than those scholars that are so frightened by the physical finitude of the planet. His great hope lies in knowledge and technical progress. Raising the welfare level of a growing population and, at the same time, saving the biosphere with the help of technical progress are, on the other hand, an insurmountable dilemma for all the scholars originating in Roegen; durable development is a nonsense: with or without technical progress, with or without free market, sooner or later resources ultimately get exhausted. Hope dies with the last resource! To move this gloomy horizon away for as long as possible, we should accept that complementarity is the norm and the basic rule, rather than the capacity for substitution! Or at least, we should avoid extremes! Not an "absurdly hard sustainability", as W. Beckerman calls it, but rather a simply hard one, as Daly does; because "strong sustainability also provides a better way of respecting the rights of future generations than does discounting" (Daly, 1995, pp. 53–54). Thus, it might be better to leave the environment in the care of other sciences, not the ones that deal with substitutions, whether they are hard or weak. This is what M. Jacobs believes: "[a]s well as providing human beings with many economic benefits, the environment has moral standing which must be taken into account in public policy, a process for which economic methods are ill-suite" (Jacobs, 1995, p. 66). That means, this problem is not fit for economists. But is it fit for ecologist economists though?

In summary and if we try to answer the question in the title, what good is this quarrel (Daly, 1997; Solow, 1997; Stiglitz, 1997) and what can be gained from it?

A first remark is that exclusivism is not decent. Sustainability does not get support from extremes; rather, equilibrium and carefulness seem to define its path.

Secondly, both parties possess frequentable and sensible areas for a kind of theory and a policy that is economically sustainable. Both the former and the latter scholars are worried for the fate of future generations. We do not know who is more concerned; some of them are worried, others are full of hope. Some see limits everywhere; others find solutions to overcome them. The "physiocratic" catastrophe-driven attitude of Roegen's disciples does not lack sensible elements altogether. When they think that some "environmental baggage" should be carried by every company, that the environment as a whole is not homogeneous but rather very specific, that everything is not sustainable, that geography is critical for development, that the process of both degradation and responsibility is both global and territorial, that biosphere implies the refusal of the standard theory of private property rights, that the eco debt must be doubled by a financial debt, and that ecological degradation is a serious and critical problem, etc., on all these points they are absolutely right. Nobody's arbitrage is needed here: these issues are common sense. However, when they invoke solutions like the supporters of degrowth do, when they believe in voluntary austerity and global solutions, but not in the saving graces of technical progress, when they blame the market and capitalism of degrading the environment and of causing inequality, they become less reasonable and less credible. And they do this in contrast with scholars who are least ardent in supporting their arguments, but who trust the creative power of man, technical progress and its self-propelling virtues.

The latter also trust science, as it has come up with solutions to big and serious problems, despite all its failings, so there are no reasons to believe that economic science would not be able to also inspire policies for the big issue of natural limitations.

6.3 Nature Versus Institutions. The Institutional "creative Destruction"

"Why nations fail" is a difficult question to answer. Greatly summarizing the big diversity of causes, it splits economists into two big groups. Some believe that geography is first of all responsible for the state of a nation. Others, however, consider that institutions should play this role. There are also middle of the road positions. The theoretical attempts on this matter are hardly new. The North–South dialogue took place based upon a similar paradigm. The successes of the New Institutional Economy in persuading that in the development process, institutions matter very much (institutions in the sense of formal and informal rules) generate counter-reactions. Climate, geographical settings, the mineral and soil richness, etc. also fight for primacy. The quarrel that is taking place today is not between the New Institutional Economy as such and all the scholars that have considered geography the main determinant of a less prosperous South. With references to such areas as well, the protagonists of this conceptual duel of our time come up with special scenarios and arguments. The triad Daron Acemoglu, Simon Johnson and James Robinson (AJR) see in the "extractive institutions" and in the "renversement de la fortune" the main causes of the partition of the world into the rich and the not so rich. Jared Diamond (2005), on the other hand, is partly associated with an excessively geographical determinism in his explanation of the different directions of development. Jeffrey Sachs (2003) tries to soothe the spirits and to impose the complementarity of causal circumstances in explaining the origins of wealth and poverty.

What novelties do Acemoglu and company bring to the table? (Acemoglu & Robinson, 2012; Acemoglu et al., 2001). Balancing the role of the three paradigms—institutional, cultural and geographical—in the making and the dynamics of the world (we would include the cultural paradigm into one of the informal institutions), they believe that only the institutional paradigm may elucidate why people live very differently on both parts of a fence sharing the same economic geography. Tropical climate may be the cause of laziness and may confirm

Montesquieu's hypothesis, but this explanation is marginal. "The geography hypothesis is not only unhelpful for explaining the origins of prosperity throughout history, and mostly incorrect in its emphasis, but also unable to account for the lay of the land we started this chapter with" (Acemoglu & Robinson, 2012, p. 54). Institutions are the ones capable of explaining everything. Institutions can be: (a) "extractive"—rules established to the benefit of the oligarchy and the metropole, but against collective prosperity (Acemoglu & Robinson, 2001); (b) "inclusive"—favourable for collective prosperity and the growth of productivity (Acemoglu & Robinson, 2012). This distinction does not seem to trace its origins to theoretical concerns. The first European colonists have made it by thinking of different strategies according to their own interests. In poor and less populated regions like the colonies, Acemoglu and the others think that they had the tendency to establish inclusive rules. On the contrary, in populated and rich-in-resources areas, European colonists considered that extractive institutions were the most suitable. After decolonization, a special phenomenon of dependency manifested itself: the inertial force of the institutions imposed by the European colonials had its say. The countries with inclusive institutions established by the European colonials kept this institutional arrangement and had a lot to gain because of it. They performed so well that, by a "renversement de la fortune" they caught up to and even surpassed formerly rich territories, but where extractive institutions were put in place! In short, today's differences in development levels are explained by the differences in strategy adopted in the past by the European colonists (David, 1985). The compensation of development differences was possible due to the permissive dynamic of inclusive institutions. Once the geography and the environment are given, they would not have permitted something like that!

On the other hand, geography can explain these differences just as well and even deterministically, Jared Diamond believes (Diamond, 2005; Diamond & Renfrew, 1997). According to him, the natural and biogeographical endowment of Eurasia explains why nations of this zone have come to oppress the African and Latino-American ones, not the other way around! Favourable biological conditions allowed them to skip steps on the road to development. Food plentiness, the division of labour, population growth, the multiplication of professions and the quickening of technological progress as well as its diffusion were steps that could be made rapidly because the foundations, namely geography, was generous! And owing to such foundations they afterwards engaged into colonial adventures. It was not for any racial or war-driven considerations, our author believes! Geography is to blame for all the successes or failures in this world! Huntington's classification (Huntington, 1998) would seemingly validate the notion that the cradle of great civilizations originated in Eurasia! This idea is not foreign to Braudel's analyses and conclusions (Braudel & Murgescu, 2002).

Geffrey Sachs argues convincingly for a balanced combination between institutions and geography, as they separately cannot explain just everything (Bloom et al., 1998; Kahn, 2015; Sachs, 2003). With a quotation from A. Smith, Sach explains and resolves this false dilemma: geography or institutions? The quote is from the institutionalist Smith, who was convinced by the importance of fundamental institutions for the dynamics of society, but who does not hesitate to show in Book I, Chap. II that the harsh geographical realities of Africa explain "the state of barbarism and poverty" of that world. In other words, both geography and institutions matter, and the direct reference to a classicist settle the matter. In accordance with Smith, Sachs denies any monocausal theory of poverty. Poor governance, cultural barriers, bad economic policies, the poverty trap, geopolitics, etc. may very well be added to the geographical position as explanatory factors of poverty (Kahn, 2015).

What can one say here?

It should be said, *firstly*, that the concern to bring a plus of clarity in this causal relation between geography and/or institutions, on the one hand, and poverty/wealth, on the other, is valuable. We must admit that the founders, both the classicists and the neoclassicists, left a generous legacy with respect to the role of institutions, unlike the legacy concerning the role of nature. But they did not disregard the latter. As we have shown, to Smith, Ricardo, Malthus and Mill geography appears as a cause for possible limitations regarding returns, food for the population, transportation conditions, international trade, accommodation or the supply of industry with raw materials. Nature was actually the Mecca of the economist for Marshall. Noteworthy is also the naturalism of these scholars, namely the philosophy of the human–environment relation. They were openly interested in finding out whether man is part of nature, a piece of it, and then he must let it be by integrating himself into it, or man is different from nature and then he must master it, change it, and not leave it free, because only he is free. If he fails to transform it, he is not a revolutionary and makes Marx angry! Only Marx!

Secondly, there is no need for scholarly argumentations to convince that geography matters. Landes said it as well, without hypocrisy and highly qualified. He believes that geography hides an “unpleasant truth”, namely, that “nature like life is unfair, unequal in its favours; further, that nature’s unfairness is not easily remedied” (Landes, 1999, pp. 4–5). The fact that institutions can be changed much more easily is mentioned, in various notes, by Acemoglu, Diamond and Sachs. If the good institutions and the good practices invoked by Smith, Veblen and North can contribute anything, this is the place to prove it, the place where nature has been unfair. The extent to which such unfairness may be compensated institutionally has been, for approximately half a century, the concern of Nobel awarded institutionalists. References to ancient history to find out the original causes of today’s poverty or wealth are at the same time audacious and risky. Both Smith and, especially, Mill did not lack subjects on the matter. Using the premise of slavery as a component of human nature, they also saw English colonization as a springboard with propelling effects superior to the colonization by other European nations. Everybody should wish to be colonized by the most civilized nation! The dialogue presented by Diamond between Yali and his Papuan friend, from *Guns, germs and steel* does not look like living a beautiful dream, but rather seems to be a deep indictment. It is not because you, white people, came to “discover” us, that we are living a good life today! You should have better left us to live in peace, seems to be the message of the aborigine.

The added food supply offered by a generous nature does not in any way explain the colonizing impulse. This subject is dangerous, we all know it. We know that it was not just Nature that was unfair, Human nature was also unjust. If one cannot make compensatory redistributions in terms of nature, one could do it in regard to Human nature. Sachs, invoking Smith, thinks that people oppressed by unfavourable natural conditions deserve to be helped. But don’t the people oppressed by a perverted Human nature who made of slavery a civilizing art also deserve to be helped? Acemoglu explores the beginnings of history and sees the colonizing impulse of implementing in the poor and poorly populated world “inclusive institutions”, a generous and inspired gesture. In the rich areas, “extractive institutions” were proper! It’s logical! To extract, there must be wealth! You cannot get blood out of a stone! That is, for the poor world colonization was an act of benevolence: rich colonists went to those places animated by charity. What a beautiful history! For those with a generous geography, woe unto them! Nature brought them uninvited guests! This historical background which one can count on to draw conclusions for the present is interesting and thrilling. But the subject is too heavy to present the "reversement de la fortune" as a credible idea for the young man in today's Saharan Africa. He also wishes, just like his ancestors analysed by Acemoglu or Diamond, to have been born in the Austrian or Italian Tirol. The reciprocal does not apply! This is also one of the reasons for today's one-way direction of emigration. It is never from Germany, Italy or Great Britain to Syria, Afghanistan or Somalia! Seeing that the "reversement de la fortune" is still to come, the poor are in search of "la fortune" in the fortune's home! History is more complicated than that, and determinism, be it geographical or institutional, does not explain it, but rather complicates it. These two circumstances do not compete with each other in explaining the state of a nation or the world. Complementarity, as it comes from the founders, has a chance to make things clear. From them and from the scholars who honoured their offer, we find out that geography is (for some) an ingrate limitation. On the chess table of Smith's world, we understand that interventions have a shot only with respect to rules and institutions. For most founders, the market is the one that intervenes. The force of Acemoglu's inclusive institutions in bringing prosperity lies with the market as well. The "inclusive institutions", he believes, have brought prosperity because they were pro-market: they encouraged the mobilization of capital and investments, industrial successes and capitalism. It remains to be seen how the market can help reach a holistic objective—collective welfare. Is it by way of individual well beings? The state is also not foreign to this scene. Jeffrey Sachs calls it, reluctantly, an actor with an important role. Also, the institutionalists of the new generation do not ignore it. But they do not place nature in opposition to institutions and they also do not subject it to any exercise of substitution with any other factor. Au contraire, they assimilate it into a new triad: technological capacity, resources and institutional agreements. And it's true, they see in the institutional arrangement the main lever to stimulate (or slow down) an economy. And their solution is not an "institutionally creative destruction", but institutional dynamics.

And there's also another problem. The examples theorized by Acemoglu or Diamond concern a world organized and quartered within fixed boundaries. Today, institutional arrangements have a strong universalist component. The North is not indifferent to what the South does and the other way around. Linked by globalization, the poor hope for an equalization. Who is to come to them with inclusive institutions? The rich states? Multinational corporations? Do they have expectations for a "renversement de la fortune"? (Sachs et al., 1998).

6.4 Marx's "Machines" and Contemporaneity

6.4.1 Substitution Supported by Perverse Dialectics

Besides the chapter "On machinery" from Ricardo's *Principles*, Marx's (1973) *Fragment on machines* is also part of the forgotten things. Actually, they are overlooked to the large detriment of all contemporaneous sustainability scholars, preoccupied with the analysis of the substitution of production factors.

In short and to the point, this *Fragment* says everything that an economist should know in order to answer the heavy questions raised by the dynamics of substitution, driven by "machines" and the technical progress. We were surprised that it was not an economist who strongly exploited the Marxist offer from this condensed text on machines. A Media specialist, Mrs. McKenzie Wark, drew our attention because of the title of her book *General intellects* (Wark, 2017), a syntagma used by Marx that the author refers to. Actually, the ideas in Marx's text may just as well offer food for thought to economists, philosophers, sociologists and more. If we consider the fact that we deal here with a technical version of Marx's position on the relation capital–labour, in counterbalance with his social position in *The Capital*, full of pathos and revolutionary projects, all the more it means that the economist, the sociologist and the historian can join together to detect the whole phenomenology of said relation.

What does the "technical" Marx tell us? He writes that in the capital–labour relation there is no doubt in regard to which one comes first! The machines (in a large sense) are obviously "products of human industry", brains and hands, and they are "the power of knowledge transposed into objects". The capital, under the form of fixed capital (a summoning of Marx's understanding of "machines"), indicates, he thinks, two things: (a) the extent to which "general social knowledge has become a direct force of production"; (b) the extent to which "the conditions of the processes of social life itself have come under the control of the general intellect and have been transformed in accordance with it" (Marx, 1973, p. 706).

The theory on the society of knowledge may well turn these words into a well-deserved motto. Just as transparent and up to date, Marx (1973) emphasizes the intimate drives of self-development, imprinted on the economy by the "machines". An important and special process of substitution takes place and, thanks to it, "the machine is the one that possesses the skill and the power instead of the worker, and it is itself virtuous". This machine in its most complex automated form becomes "a moving power that moves itself"! In relation to the machine, workers become "conscious connections"; this means that the machine does not mediate: it replaces. Man's work is regulated and becomes an abstraction thanks to the machine, not the other way around. Science, which Marx does not overlook for a second, has the last word in establishing the contents of the relation worker-machine! These are his words: "The science which compels the inanimate limbs of the machinery, by their construction, to act purposefully, as an automaton, does not exist in the worker's consciousness, but rather acts upon him through the machine as an alien power, as the power of the machine itself" (Marx, 1973, pp. 614–615).

This is admittedly a text of exemplary reasonableness and honesty! Neutral and equidistant in relation to the two terms of this equation, capital and labour, Marx the scientist tells us how the general intellect, science and technical progress infuse not only the economy, but also the social practices. In other words, the "science transposed into objects" imprints on the economy a sort of technical laissez-faire and makes it become "a moving power that moves itself".

Is this technical determinism? Maybe it is, but Marx only wrote these lines on the matter of substitution. Here, in these lines, substitution is simple, self-explanatory, and in addition doable with the aid and the approval of science. However, substitution becomes difficult and conflictual in the texts that Marx writes in his role of professional revolutionary. In *The capital* he separates the levels, compelled by his adopted methodology. And here, we can see that the social representation of the relation capital–labour seems to be much more important to him. As such, he empties the relation of its technical and neutral content, and fills it with reasons for conflict. He links capital to the "bourgeois", with his laziness, greed and unchecked egoism, and labour to the worker, humiliated and compelled to work hard to allow the former to live without working. Twisted in such a manner, the relation could not have followed any joyful path nor conclude with a happy ending. On the contrary, in this case, it is no longer important which came first. It is important to remember that capital, initially based on theft, deserves to be denied of what is in it and against it—materialized human labour. At the end of the road, the capital would abort its mother, who has become a bourgeois exploiting tool. The technical role of a "driving force that moves by itself" is not completely forgotten. It appears in the mechanism of transforming simple labour into complex labour. And it does not appear in just any form, but to the apparent benefit of the working class. Without any doubts with respect to the role of machines in shortening the working hours, Marx writes that "capital here - quite unintentionally - reduces human labour, expenditure of energy, to a minimum. This will redound to the benefit of emancipated labour, and is the condition of its emancipation" (Marx, 1973, p. 620).

With his specific dialectics, Marx claims a lot of things. *Firstly*, it is not the intention of the capitalist to reduce the working hours by introducing more performant machines. But he does it "without intention"! We learn from the analysis of the absolute and relative surplus value that the master of capital is only interested in increasing the physical and intellectual potential of the worker. *Secondly*, "energy expenditure" does not refer to any motor form of energy consumed in the production process. Marx speaks of human, physical and/or intellectual energy. He was not so naive as to not know that the economy of his time was fuelled by coal, wood or oil. He knew, but that was not what he meant. This is why we believe that the accusation of Mrs. McKenzie that "Marx did not understand that replacing human energy with that of fossil fuels occupies the centre stage in capitalism's development model" is missing the mark (McKenzie, 2018, p. 26). Marx knew, but he was not primarily interested in that. A technical mode of production, founded and based on a certain resource, did not concern him. His goal was the capitalist mode of production, its "metabolism"; a system in which two "toys", two competing forces—the means of production and the social relations of production—define the pace but also the moment when capitalism hits a wall, implodes and dies. Self-development, that "moving force that moves by itself" from the *Fragment*, concerns now only one component—the forces of production. And in this area, revolutionary things happen. A "devilish" force captures science, enslaves it and puts it to work. As a result, the forces of production gain a head start. When the distance is too large, the relations of production become "cuffs", an insurmountable obstacle. The situation is not resolved through substitution, but conflict. Labour, that is stored in capital, gets its revenge. Capitalism fails and, concurrently, the child, cleansed of his bourgeois sins, gets back to his natural mother. The one that works, the proletarian, becomes the master of what is rightfully his. And he deserves the capital as well. That's where his labour is! The spent entrepreneurial energy, the taken risk, the managerial effort, none of them matter!

On the path of the *Capital*, but especially at the end of it, the technical substitution of labour with capital, determined by reasons of ease of work, is lacking logic. There is no longer a "machine"—mediated complementarity between the two main factors of production. They are incompatible because their masters are "objectively" enemies. Marx was totally foreign to the bourgeois arrangement. His few reflections bearing a scientific tendency are lost in the ideological shell in which they are embedded. Any "proletarian-bourgeois" agreement is excluded. Substitution is out of the question. Capital, the "devil", must be removed.

Only when he speaks of communism does Marx's face brighten as it glimpses a happier face of the capital–labour relation. In fact, capital "disappears" and becomes a "means of production". And this happens as a result of a party directive. For the people who lived through communism or have studied Marxist political economy, a party directive means that the Party knows exactly what the means of production and the workforce should look like. There was no problem of substitution! The Party orders that labour "marries" capital! Everybody has to work, irrespective of how many systems of machines the country possesses. That is, capital no longer threatens anyone but it keeps everyone busy. Only angels may stay unemployed!

6.4.2 From Marx's Machines to Its Highness, the GENERAL INTELLECT. The Multifactor Productivity

At least four aspects are revealed from Marx's dialectics on substitution.

Firstly, Marx's "machines" have multiplied, have become diversified and complicated, not because of self-development itself, but rather under the umbrella and with the participation of science and knowledge, whose fruits they are. *Secondly*, technical progress infuses not only the economy, but also the society and life in general. *Thirdly*, substitution does not only happen between labour and machines, but also between the diverse and multiple versions of labour and capital. *Not lastly*, the "general intellect" decreases more and more the material component of the substitution processes in favour of the nonmaterial, namely the grey matter. What do we mean by that?

Smith was talking about the way good habits, under the form of civic externalities, are diffused within one generation, and also between generations. Marx's "machines" follow a similar pattern. They continue to be a result of labour, of human capital. This is the starting point. Before becoming "itself virtuous", the machine gets its "skill" directly from science, by means of the human capital employed, and for a long time, it drives a cumulative process in which today's new idea is a trump card for tomorrow's new idea, without in-between interruptions. It is what R. Lucas (1988) and other authors like Angel de Fuente and Antonio Ciccone (2003) call "rate effect". In other words, a stock of human capital influences productivity on the entire range of factors. Embedded in their very essence, the influence can be felt both today, as well as in the future. It's difficult to dissociate and measure, in time, the influence of human capital. It goes through generations. How is one to establish the extension of the private property right over an idea? As we have said, a cumulative process is here at work and it is difficult to unravel the "lump" of ideas in order to identify the contribution of each and every creator. Actually, the impregnation with knowledge of human capital, and through it, of all the other factors allows the syntagma "productive multiplicator", used to show that technology cannot be against any domain, including nature, and that by using this multiplying factor one can obtain higher results in using natural resources and labour, facilitating technological innovation, investments in non-polluting infrastructure, the improvement of skills and professional competence in view of creating new jobs (Smith et al., 2010). In short, a "productive multiplicator" offers the chance for relaxation, for the uncoupling of growth from environmental pressure precisely thanks to the way in which it operates concomitantly on both flanks. This is the context in which we should appreciate that Marx's "productive collective worker" is today's engineer and researcher, who does not take part in any course on sustainability and substitution of factors, but have written scholarly works. Under their signature and through them, science and technology talk about the capture and storage of carbon dioxide, photovoltaic systems and "the way to the Sun", nanotechnology or the urban mining of recovered metals, the substitution of metal by plastic materials or of petroleum by ethanol, etc. He is the engineer who gives life to technical changes à la Marx, whose involvement determines the growth of the organic composition of capital in the sense envisioned by Ricardo and especially Marx (Kurz, 2010). He is the main character, but not a solitary one. The "general intellect", the general social knowledge, places him beside the sociologist, biologist, political thinker, geographer, physicist, etc. This way, he finds out whether his technical solution is also admitted, whether ethanol may substitute oil, but only after having fed the people first, and only afterwards machines.

Marx's "machines" associate nowadays capital and labour to a whole bunch of other factors who give a helping hand to general business. Management, practical experience, the openness to take risks, intelligence, "science as a production factor", etc., take part effectively and not as in the time of the scholar who gave his benediction to this idea. If this is no longer just a "productive triad", but rather a very large range of factors, isn't it natural to be concerned with finding versions of substitution and optimal combinations thereof to maximize results? The scholars who made sense of these judgements à la marge considered the mathematical calculations employed by this process to be even charming. And they are not the only people who are trained to determine substitution coefficients and to discover the extent to which capital and information, management and work, risk and praxis are combined, but also separated, etc. Modern accountancy has challenges when it is to find out the percentages of participation to costs. Today's versions of the theory of transaction costs try to identify and to quantify the percentages of such costs that evade accountancy, which are "present without being visible"! The Heckscher-Ohlin-Samuelson theorem tries to recognize a message in the same paradigm: to find out what factors concur to the production of a good meant for commerce and to focus on such products that use as much as possible of the plentiful factor and as little as possible of the rare factor. In short, use as much super-qualified work as possible and the least amount of matter. Substitute, if you can, up to 100%, any factor with grey matter, this seems to be the command! Sell the labour and keep the capital. Keep Solow's "technical capacity of production", the "man-made capital"! And, by the way! What is left of the famous "residue" from his model in this context in which, over all of them, the master is His Highness, the general intellect? The idea that Solow does not deal with its origin, shrouding it in a mystery, as if it came down from the Heavens and its usage in production leads to an exogenic growth, is one of the most short-sighted. Actually, one doesn't even have to deal especially with the origin of technical progress in a growth model. The same goes with the origin of labour or nature. Even the scholars on whose thoughts he found support did not do it. And for Rostow, technical progress is a technical sum of scientific views with a great power of dissemination in all sectors of activity. This is how new industries appear, using new capital and generating new possibilities of development. This is how the famous stages are covered and the wished-for self-development is secured. The replies to the idea delivered by Grossman and Krueger (1993) do not reduce its force of suggestion. In a similar manner, in 1939, Schumpeter was explaining economic cycles by means of technical progress. The one who gives growth a path is technical innovation generated in waves. By propagating technical progress, a wave prepares the next. The relation is one of self-maintenance: progress supports growth, growth serves as a springboard for progress. This means that progress is naturally endogenous! And then comes Solow and expects technical progress to fall from the Heavens and cause exogenous growth! Come on!

"Man-made capital" is, for both Solow and Stiglitz, a stock of knowledge, equipment and competences. It is "general social knowledge that became a direct production force", as Marx (1990) put it. With this concept in mind, both scholars conceive of possible combinations to maintain or even maximize utilities. And they try to reach the aim, considering the great ability of technical progress to "contaminate" everything, both material and nonmaterial forces, nature included. Criticisms of the "residue" determined Stiglitz to integrate in the 1970s the environment into the well-known production function Cobb–Douglas. He thought that the environment would reduce constraints precisely thanks to its infusion with technical progress. Soft sustainability was conceived this way: by self-development being driven by technical progress and by the growth of efficiency of the natural factor, and not just anyway, but to a larger extent than the growth rate of the population, and in addition by responding to warnings of the type delivered, for instance, by Vincent Martinet (2012) or Martinet and Rotillon (2007). The message transmitted by the models of the two Nobel Prize winners is not translated into exalted enthusiasm. Actually, it is a reasonable hope for endurance based on the trust in man's creative force. It is the confidence in Marx's "machines", whom they do not quote, even though he validates their thoughts. We are talking about the Marx of the *Fragment* and not the one of the *Capital*. Are they not aware that nature also has its own limits and that it may also refuse the impregnation with science? We doubt that they don't know. They do know, just like everybody else, that lacking attention with respect to nature may be conducive to the situation in which there is nothing left to infuse and impregnate with technical progress and advanced science. But if against such a background, they offer us a glimpse of hope, why should we waste it? What reasons do we have to think that, faced with ecological disasters, humankind would be incapable of producing another Einstein, Tesla, Turing, Edison or Watt?

This subject leads to a curious paradox. Science has become a redoubtable force of production, as Marx predicted. Its effects are dissipated upstream and downstream of any idea that assists growth. However, by following such a path, the "general intellect" invalidates the Smithian and Marxist conception of productive and non-productive labour. Why so? Once consumed, the general intellect does not turn into anything palpable. The scholars who nowadays extend Turing's initiatives have issues in decisively marking the property boundaries of their own work. Their work is knowledge: once "produced", it invades. We do not have a special crush on the syntagma "society of knowledge". We think that it also admits its own opposite, a society of ignorance, which was left behind in the past. If we make references to the past on this matter, we believe we may have some problems. The great cathedrals as symbols of an exceptionally creative force, do not justify and do not contribute to the definition of the "dark Middle Ages". The same applies for the Aztec temples and cities as well as Egyptian pyramids. However, we accept said syntagma as a form of labour: invasive, but not without certain meaning. We accept that there is nothing today without being included in knowledge. Today's civilized world, Popper's open society, is open both to democratic values and to ideas and money. Everything includes lots of brains and little physical matter; and everything is expressed financially. In other words, it is a paradox to talk nowadays of a threat and a collapse of renewable resources, when the proportion of intelligence in the national wealth of countries that formulate the issue in those terms is so high. If values are extracted to such a large extent from digital labour, why are we so afraid of the collapse of something with such an abrupt drop of share in the structure of wealth? (Scholz, 2012). With or without the accord of Solow or Daly, Marx's new "machines" are served today by the new human subjects who accomplish an unprecedented substitutional revolution. When working almost exclusively with people's brains, researchers and engineers offer us the display of such substitutions that we have never imagined. Such substitutions are not between factors, but rather within them; man, technologies, resources, praxis, management, etc. It's obvious we are not about to eat IT as the first course and we also will not make a roof out of artificial intelligence. But having bad dreams for the fear that we might sleep in the open air unless we speak Roegen's language is not a solution either.

6.4.3 Sustainability in the Era of the Technological Unemployed

The great substitution committed to be produced by the general intellect responds, in essence, to the same commands as the "small substitution": the facilitation of work and the scarcity of natural resources. From a technical–economic point of view, the situation looks promising. But, whatever it may be called, weak or strong, substitution cannot be removed from its primary essence, the capital–labour relation. The replacement of labour with capital has not been a mechanical act, even though one of the equation terms belongs to the inanimate, material world. On the contrary, it has borrowed something from the tension of the hiring process. Obtaining a job was and remains a justified desire and need, socially and economically, of any individual. However, a sophisticated and ultra-productive machine, ready to take over an increasing number of tasks from the job description, has appeared as a possible threat. The fact that it might ease one's work did not compensate for the fear of losing one's job (work). The refusal of what seemed normal to happen looked rational to the worker. It is interesting that, if we consider things historically, it's not only the replaceable person that resisted the process. Reasons beyond the worker-employer tension have generated paradoxes here as well. The typical case is the cotton mills in colonized India. There was enough capital to replace hard and back-breaking labour. But things did not happen this way. Using marginal analysis, the owners of capital found out that the excess demand could be honoured not by using an additional capital unit, but rather by employing cheaper workforce. On the other side, the wish to divide work and to secure sustenance for as many people as possible, including women and children, was translated into the refusal of the new "machines". Overall, the work easement reason did not influence the decision. Profit interests along with the desire to have a job, even in humiliating conditions, won. Between an expensive machine and a cheap worker, the choice was clear (Landes, 1999).

Today's lohn system is a modern formula, with a new name, but based on a logic similar to that which operated in the old Indian cotton spinning mills. What did the process of "creative destruction" do in the meantime? It worked; however, hundreds of years after the time of the Indian cotton mills. Davos 2016 brings up the same issue, at other levels and in different nuances, without leaving aside the main problem. In short, digitalization eased labour at unimaginable scales, but it also threatens to cause unemployment. The digital unemployed is the new category of people seduced and defeated by sophisticated "machines". *The Future of Employment* (Frey & Osborne, 2013) and, more recently, *Google Archipelago: The Digital Gulag and the Simulation of Freedom* (Rectenwald, 2019) are just two examples among hundreds, revealing the fact that the "digital tyranny" causes serious and contradictory mutations. While robots take over whole branches of traditional industries as well as services of income distribution, state surveillance, health, education, duties and taxes, police, etc., traditional crafts disappear and others, brand new, make their appearance. Neither Quesnay nor Smith or Marx heard about supervisors in communications and information or specialists in face recognition technology or about the way in which large private corporations are able to substitute governments in important tasks by means of the "Google Marxism"! Intelligent robots replace labour in an overwhelming proportion of about 80%. This is indeed a significant substitution. What remains beyond this percentage is increasingly insignificant.

What do we do with the "useless" ones, the prisoners of digital tyranny? There are already a lot of studies on the matter. From the very beginning, the theory of sustainable development has drawn attention to the deterioration of the relationship between physical and intellectual labour. However, it has done it from the perspective of people's health, caring for the sedentary IT worker, who is super-intelligent but with health problems because he only moves his mind. Useless people were never a problem; now they are. The study *Does capitalism have a future?* (Wallerstein & Collins, 2014) asks intriguing questions on the subject, with bold attempts at answering them. Experiencing trepidations similar to those in Schumpeter's *Capitalism, socialism and democracy*, the authors theoretically test the solution of a socialist reorganization of production and distribution. This notion does not seem reasonable to them, but the alternative to socialism, capitalism, is no longer interesting to them either; the ecological capitalist failures no longer recommend it. A "different organization" seems more to their liking, a mishmash, painted green on the outside and red on the inside. The idea of financing ventures, in extremis, to employ the workers coming from industries, is also analysed. Moving industries to less developed countries is under discussion. One perspective is to see the government as a main investor and employer! The idea of us all becoming capitalist is also not bad! But beyond fantasies and coquetries, a powerful idea emerges. It refers to the type of school to cultivate, a school that should face not only a radically new demand, but also to the pace at which it is demanded. This school must understand that the reaction time is very short, here and in the case of other resources that need replacement. It is a school of adaptive anticipation with maximum reaction speed. The solution is not specialization, Braudel claims, but the generalist type, potentially adaptable. We believe he is right. The generalist will orient himself towards anything, including the newly appeared services in the ecological sector. Environmental activists will not be alumni of physical mechanics. But however the work market will look like in the digital era, people will work seriously. The perspective of robotization may be a great disillusion for the people who, inspired by the atmosphere of an eventual stationary state à la Mill, would think that poetry, contemplation, sweet daydreaming and scholarly laziness would be the main jobs in the time in which Marx's "machines" and the "general intellect" take centre stage. They would find out that they must work. It's true, they must work in a different manner: sustainably and resiliently schooled, with diplomas of excellence for the future. They would bolster the middle class by raising its living standard. Smith's merchant, loan shark or "shoemaker" are not the pieces of the game. "Geometrically" speaking, the middle class would not be in the middle anymore; it would be, and it already is at the top. It has not vanished. It floods the field of services and produces preponderantly nonmaterial wealth in the domain of the "superior cognitive processes". What happens in the material domain? It exists, but it is served by fewer and fewer people. But anyway, it cannot disappear. The wolf cannot survive without the sheep.

6.5 Is There Need for a New Science to Achieve Strong Sustainability?

Kuhn and Lakatos are two important scholars with instructive works about the battle between the two paradigms and the possibility that, after this struggle, a new one is born, defining another science. The invitation to relate to the intellectual adventure of those scholars who baptised it, to find out the significance of the very vocal argumentative exercise of those who want to promote Ecological Economy as a science, seems acceptable to us. We have to find somewhere, and not just anywhere, some criteria with respect to which we should find out what separates the old science from the new science and what the definition of a science should ultimately sound like.

It's beyond doubt that a science has an object, a method and a proper notional and epistemological basis. Before the definition of its object becomes final, there is a long historical process of clarification and conceptual decantation. After this happens, there follows an awaited and deserved process of classicization. In that stage, the mentors, the apostles and their writings and the creators of systems stand out. They set the tone and fix the method, define the specificity of the language and draw the boundaries between which one may or may not interfere with other sciences. After its foundation, science enters a permanent dynamic. Methodological revolutions may take place within it; its object is always updated, by receiving from and giving, in its turn, to the sciences to which it is organically linked.

Economic science has been no exception concerning this process. It has been recognized and respected as a science. During crises, it is accused of all things, but it overcomes such allegations and remains a science. Is Ecological economics in a similar situation? Is it a science? Let us try to give an answer in several steps!

Firstly, what is the necessity of its appearance? A basic answer would be: ecological economics has appeared against the background of the dissatisfactions of its godfathers with respect to the inability of standard economic science to satisfactorily solve the issues occurring from the economy's relations to nature. Here, the standard science is largely identified with its current neoclassical version as a form of manifestation. In this context, it is refused and distance is also taken from the gentle, neoclassical friendly offer of Environmental Economics or Resource Economics. Frustrations are not few. The lack of consideration for the value of natural elements and their correlation to their total economic value is one of the charges (Douai & Plumecocq, 2017). Specific to this idea is the criticism of the non-inclusion of the depreciation of natural capital into national accounts. A concept like "Throughput", indicating a flow of energy-matter, with inputs and outputs, considered very necessary for a real result analysis, does not find a place if the ecological factors are left aside (Daly, 2008; Goodland et al., 1992). The problem-free dynamics of an economy setup to continuously grow is not a perplexity, but a defiance of a reality that has observable limits. The obsession with limits and, on this matter, prudence: these are great ideas that must find their place in the theoretical body of the new science (Daly, 1990; Passet, 1996). This is not just about limits and forethought, but the new science must make its appearance to talk about hard vs weak sustainability, to argue in favour of complementarity in relation to the substitution of factors, to concern oneself in a special way with nature as long as neither the classicists nor especially the neoclassicists really fell for this subject matter, and why not, to lend a helping hand to the excessive dematerialization of life. In short, a Political Economy of sustainability or an Ecological Economy whose next of kin is Roegen's Bio-economy!

Then, who is to open its path? We believe that three circumstances smoothed the birth of the notion about this newly professed science.

The first moment is called the Roegen moment. His analysis of planet limitations from the position of a scientist and his arguments in favour of Bio-economy have had strong reverberations. Today, Bio-economy has taken over the stage and seems to be the signal word everywhere. However, is Nicolae Georgescu-Roegen the mentor with auroral insight of the new science if he is quoted both by the standard science and by Environmental Economics as well as by Resource Economics? In the preface to the book The Origins of Ecological Economics, written by Kozo Mayumi, a former student of Roegen, Joan Martinez-Alier claims that "what Georgescu-Roegen (1906–1994) called 'bioeconomics' has come to be called 'ecological economics'" (Joan Martinez-Alier, apud Mayumi, 2001, p.ix). On the other hand, another brilliant disciple of Roegen, Daly, acknowledges the "stationary state" while in Roegen only the degrowth is admitted. We should also add that the discourse on degrowth only revolves around Roegen! We are not bothered that Roegen inspires several sciences. We are interested to know whether the reputed economist may be considered the forerunner of the new science that scholars are talking about. And we ascertain that there is no direct answer at hand, even though the issue of resource limitations has also constituted an obsession of ecologist economists. And then, where are we to find the founders? Is it among economists with proclivities towards biology or among non-economists with a passion for economy? Can we use Hayek at the former rubric? Or the medic Quesnay at the latter? This is a tough task! Actually, "historical" searches do not go that way. They go towards the initiators Herman Daly, economist, and Robert Costanza, architect, urban planner, and doctor in the science of the environment, together with the participants at the first symposium on these matters, in Sweden 1982, most of whom are ecologists. Five years later, in 1987, Joan

Martinez-Alier (1990), an economist, published *Ecological economics: Energy, environment and society*, subscribing to the founding line. The publication in 1989 of the journal Ecological Economics confirmed that this movement is serious. Once started, the movement attracted unanticipated energies. A moment of glory was considered the granting of the first doctor's degree in Ecological Economics in the world to Sigrid Stagl in 1999. We can only wait for the classicization of the moment of debut!

A *second* circumstance that had a say in the appearance of Ecological Economics and its aspiration to the status of science is related to the economists and their pale sentiment of belonging to a science that they should defend from the inside as well as from the outside. They should protect it from the intrusiveness of daily life and from verbosity. However, there is no other craft that defends its territory less or that minimizes and ridicules its own role more. Galbraith plays with this matter and claims that "Leading active members of today's economics profession... have formed themselves into a kind of Politburo for correct economic thinking. As a general rule (...) this has placed them on the wrong side of every important policy issue, and not just recently but for decades. They predict disaster where none occurs" (Galbraith, 2012, p. 95). The quotation does not end here. There follows a number of sentences just as "flattering" for the profession that Galbraith himself belongs to. Closer to us, Mr. Krugman writes about the economists who dare take on the cloth of deceiving prophets: "Few economists saw our current crisis coming, but this predictive failure was the least of the field's problems. More important was the profession's blindness to the very possibility of catastrophic failures in a market economy" (Krugman, 2009). And such examples may go on to build an inside view on the "inability of this profession" regarding many issues.

In a field thus prepared, what can the non-economists claim? They claim, for instance, using the pen of Richard Heinberg, whose work we have mentioned before, a journalist and educator on the theme of energy, that "economists say silly things", that in their capacity as "economic experts with a tunnel vision" they baffle the influencers; that "[w]hen Smith imagined the economy of the future, he foresaw one comprised of shopkeepers, artisans, small factories, and trade via sailing ships, because those were his customary terms of reference"! These quotes are from the book *The End of Growth* (Heinberg, 2012) of said author, pages 337, 66, and 507. We do not know of such future projections of our classicist. We do know, however, that he filled the space of his time with strong ideas which, both in his time and nowadays, mean science. The problem is that, on a path unhindered by anyone and anything, Heinberg (2012) also makes the necessary proposals for the way that post-growth economic science should look like. He proposes a revision of economic theory to take into account ecological limits and the destruction of the brotherhood of conventional economists. Richard Heinberg's post-growth New Science is a combination of Roegen, Daly, the partisans of degrowth, the futurist Hazel Henderson and, how could it be missed?!, the cooperative system of the "alternative economists" with all their Fourierist utopia. Who should come with proposals for a new economic science if such an issue were raised? It is, we would claim, the economists' business. It's just that, caught by the morbus of newness at any price, even "alternative" economists talk funny things. For instance, after specifying, as a preamble, that Ecological Economics is a field of interdisciplinary study, the ecologist economist Peter Söderbaum invites to such study anyone who feels concerned for the natural environment! This means, actually, that "even a neo-classical environmental economist can refer to her- or himself as an ecological economist, if she or he so prefers" (Söderbaum, 2000, p. 20). Everybody may come in, but only those who love nature stay- this is the "alternative" formula for becoming a sustainable and resilient ecological economist. Was corrosive daily life so trivializing to a science that opened itself too much to the common domain?

Then, finally, a *third* circumstance that propelled the drives of inventing a whole new science refers to the domain of facts. The 1970s and the following years have brought an acute awareness of the serious environmental issues and their connection to a science that did not give the impression of being willing to assimilate these new realities into its analyses. The reports of the Club of Rome, international conferences, some under the aegis of the UN, the Brundtland Report, etc. have created the proper context. Roegen ignited the spark and Malthus, Mill and all the stationarists who considered resource limitations reasons for pessimism were taken back out into the light of day.

Against the background of the vivid and spirited disputes on issues like the environment and the limits of resources, the scholars with little confidence in the theoretical power of the standard science produced a small schism. They broke with the mother science, started to use its name under quotation marks, and after having become the "alternative", started to use new names for themselves and to invent specific notions and instruments. The Economics of degrowth, the Economics of the environment, the Economics of resources and the Ecological Economics are the main paths which the heretical children followed.

The path with the most solid claims to the status of a real science is Ecological Economics. And how does it actually try to shape its profile and raison d'être? First of all, by defining its object: "*Ecological economics* is a new *transdisciplinary* field of study that addresses the relationships between ecosystems and economic systems in the broadest sense", founders Costanza et al., (1991, p. 3) claim. Why do they deal with such things? Because these problems, very important and fresh, "are not well covered by any existing scientific discipline" (Costanza et al., 1991, p. 3). Is there a method for such a thing? Of course there is: "Ecological economics will use the tools of conventional economics and ecology as appropriate." (Costanza et al., 1991, p. 3). This would have been suitable as a logical proposition unless in the same text they hadn't made the specification that "Ecological economics (EE) differs from both conventional economics and conventional ecology" (Costanza et al., 1991, p. 3). We differ, but we steal your work tools! It is well-known: if there is no difference in method, there's nothing! Situated on the orbit of institutional economics, the rebelled daughter cannot break away from the core because it lacks an original method. In the same manner, Roegen's bioeconomics places its object of study at the boundaries between economics and biology. This means that they are similar. What differentiates them, Mauro Bonainti (D'Alisa et al., 2014) claims, concerns their starting premises. Bio-economics denies development, even in its sustainable form; it only accepts degrowth. Ecological economics admits at least the equilibrium of the stationary state, not just degrowth (Daly & Farley, 2011). Transdisciplinarity seems to be the keyword for the methodological shape of Ecological Economics. And Constanza tells us that this does not amount to its setting just between economics and ecology, but also "it's a bridge across not only ecology and economics but also psychology, anthropology, archaeology, and history" (Costanza, 2010). And so that the fog becomes even thicker, we find out that "when we say that ecological economics is transdisciplinary, we do not simply mean that it is concerned with economic and ecological phenomena and draws on the disciplines of economics and ecology. It is and it does, but more is involved. (…) It requires a common perspective that 'transcends' those that are standard in the two disciplines" (Common & Stagl, 2005, p. 5). And, in order for us to stay in the mist, we are also told that this is not about combinations or permutations. Let it be clear for us: "… ecological economics is not an alternative to any of the existing disciplines" (Costanza et al., 2015, p. 88).

On its path both self-made and transdisciplinary, Ecological Economics sets up to solve three large issues: "allocation, distribution, and scale" (Costanza et al., 2015, p. 90). The ratio would make the difference to standard economics, a sort of flow and swing of matter and energy from man to the environment and vice versa! Where is production? Who attends to it? Anyway, Ecological Economics does not. Rather, it extends its job description to the legislative process, because, we find out, this is how it's supposed to be seen, "as a social and political science (not an extension of natural science)" (Söderbaum, 2007, p. 209). The proclivity of the economic ecologists towards ideology is openly declared and supported by the observation that "the choice of a particular paradigm is partly a political and ideological choice and not only a matter of 'good science'" (Söderbaum, 2007, p. 221). The perspective is further developed in another article by the same author (Söderbaum, 2011).

Every new method requires concepts. The new science does not have a lot of semantic innovations and proper concepts to it. It mostly works with the ones presented in the analyses on sustainability, durable development and degrowth. Natural capital is at the heart of its system of notions. Herman Daly is fond of the term "throughput". Ecological complexity, irreversibility, precaution, intergenerational equity, optimal macroeconomic scale, entropy, energy flow, complementarity, physical accountancy, the updating of the future, etc. are the keywords by which the language of ecological economists comes to life.

In summary, what is there to be said?

Any attempt to create more knowledge, irrespective of the field, deserves to be acknowledged. From this point of view, the controversy engaged by Ecological Economics on resources, on the possible type of growth and on the relation between the economy and the environment has in itself the role of keeping alive an agenda of pressing issues. However, we have serious hesitations in regard to its desired position and its claim to the status of science. If there is anything happening to Ecological Economics on this path, it may be well described by the syntagma "in the course of". It is "in the course of" getting a clear object in the large web of relations of the interdisciplinary domain in which its analyses fit. It is "in course of" finding its still diffused and borrowed methodological arsenal. It is "in course" of forming an opinion as united as possible of the scholars who support it and whom, instead of giving it a shape, bathe it either in the waters of a discipline or a school, a paradigm, or a policy, by making it similar to the science of sustainability or durability. A matured science is unlike anything else; it is simply science. It also has a problem with the scholars that serve it directly. Most of them are not economists. Inasmuch as you try to gain autonomy from a certain science, in this case the science of Economics, whose judgements you doubt, you cannot turn into something that belongs to another sphere of knowledge. You cannot literally become something else. When you fight against standard economics, you cannot replace it with chemistry, biology or physics. You come up with another economics, yet it is still economics. Economics cannot claim the exclusive right of access to the knowledge of economic phenomena. It is well-known, and its history has shown us that it has been supported by history, biology, sociology, mathematics, etc. But it has never worn the cloth of a hybrid when it approached subject matters that traditionally did not belong to it. From the position of the unitary science of economics, it also approached the new theory of the consumer and the theory of Public Choice. Now, it is invited to wear this garment. However, as long as it borrows its methods from standard economics, Ecological Economics should still be called, basically, economic science. But it is not by the will of the scholars who make it live, openly or subliminally. It is neither economics nor ecology. It is something "in the course of", hard to mark, resulting from the fact that economics as a science has left itself too much invaded by verboseness, and lately by bio- and eco-related verboseness.

It is not new that in the dynamics of the science of economics there occur heresies, be they doctrinal or methodological. Marginalism is a case of breaking with a secular tradition, coming from a new method and a new point of support—the subjective theory of value. But Marginalism, with its three founding minds and with the splitting of economists into "mathematicians" and "men of letters", is not a different science. The German historical school, with its different insights and methods, was in its turn a break with tradition, but not with science. Examples may go on. It has been claimed that after Roegen, bio-economics is everywhere, in every talk and analysis. Ecological Economics will turn out the same, namely everywhere. It's just that this "everywhere" does not include heavy names of the science of economics. Have you seen Wicksell, Bawerk, Walras, Keynes (Keynes the man of mechanisms, not Keynes the doctrinal thinker), Hayek from *Prices and production*, Mises from the *Theory of cycles*, etc. included in the analyses of Ecological Economics?

Neoclassicism is treated in bulk, as a whole, and blamed for its sympathy for the market. Solow and Stiglitz introduced the environment into their analyses, but when they are mixed in the general soup of neoclassicism, they cannot be identified as vocal names. With his inverted U curve applied to the environment, Kuznets did not claim either to have knocked the door of a different science. The predilection for developments in the legislative field and the thinning of the importance of the positive level are a sign that Ecological Economics also lacks a hard core. It is true that Herman Daly opened a line of analytical rigour and fundamental questions. Too few others follow his lead. And when they do, from their position as non-economists, they do no service to the science of economics. By criticizing its simplifying technicism and filling its spaces with science learnt from Wikipedia, it does not exalt it; quite the contrary.

In total, detachment from the whole on the grounds that, at its base, there is no true and appropriate knowledge for the present moment is neither a necessity nor a success. It is strange to see how a traditional science is left to deal with weak sustainability while for hard sustainability the necessity for a new science is brought up! Not even Einstein, with his complete novelty, claimed the status of a new science. His relativity fell in the field of Physics. But for strong substitution, we necessarily need an Ecological Economy! In relation to the need to coagulate efforts and to consolidate the science of economics, this is an idea as contrary to the spirit of sustainability as it gets. If we consider this matter from the historical perspective, with the losing and centrifugal attempt of breaking with the whole, it is just one example that competes with it, with devastating effects in the field of knowledge. We are talking about the breaking of Political Economics into two parts during the Communist experiment. This separation produced the Bourgeois political economics, on the one hand, and the Socialist political economics, on the other. The first was considered vulgar and without horizon! The second was deemed as pioneering and suitable to prescribe and to promote unbeatable truths! The lack of inflation, unemployment and pollution in socialism was part of such truths. It seems curious, but the Socialist manuals of political economics were promoting the notion that there is no technical material base in general, but only a socialist one and a capitalist one. In other words, turning machines, drills, etc. were divided into capitalist and socialist! And, also in other words, the socialist technical material base (the socialist capital) could not cause pollution; it was forbidden! This was so until Chernobyl cancelled the operating criterion in this weird classification.

The comparison to Ecological Economics is a stretch. But we have used it to suggest that this splitting, unsupported by rational criteria, is conducive to extravaganza. In any case, it is unsustainable, particularly when the part that is called in the spotlight and intensely exploited is legislation, the "ideological and political choices" of the supporters of such discourse, thus influencing the forming of economic paradigms, as Peter Soderbaun justly claims.

6.6 Sustainability and the World of Animals

Concerned with whether economics was foggy or not, or filled with "rebound effects", the theory of sustainable development unpardonably minimizes the role of unreasoning beings in this formula. Domesticated or wild, animals belong to Nature, define it and give it colour, content, and life. In short, what would Nature be without its animals?

"The Mecca of the economist is economic biology rather than economic dynamics", A. Marshall was cogitating, without considering animals in particular (Marshall, apud Pribram, 1986, p. 303). He found it more important to know for how long the coal reserves of his country were meant to endure. Other neoclassicists and classicists do not excel on this subject matter either. From the examples of the physiocrats, the perfect management of anthills or beehives, through J. S. Mill's pig, an etalon of inferiority in which nature took form, the decline is considerable. Only someone like Jeremy Bentham saved the appearances by extending the principles of utilitarianism and including animals in the moral objective of general wellbeing. In addition, he asked a Shakespearian question, in order to stir the minds of his time: "The question is not *Can they reason?* or *Can they Talk?* but *Can they suffer?*" (Bentham, 1823, p. 144). It's just that the minds of his time, just like those of ours, in respect to the tribe of economists, were neither sensitized nor compelled to think. Revoltingly few studies include animals in their environmental studies, even though we share the environment with them, as well as with trees, waters, petroleum, etc. The studies that do see to the "animal welfare" are rare flowers (Norwood & Lusk, 2011), which are cultivated in the same world of ideas as Bentham, trying to answer the obsessive question whether or not animals have awareness and feel pain! Animal lovers, which are many and have big hearts, do make up for this, on the individual or institutional level.

The relationship human-animal and animal-human or nature does not belong together with the analyses on the environment, resources, substitution, etc. However, this field does not lack a minimum paradigm. What happens with the life of animals is a question whose answer may be found at the confluence of the societal mode of organization and individualism as a behavioural norm, in the relationship between the free market, hedonist principles and the extension of moral principles to the animal world, in the connection between the culture of rules and the lack of responsibility or concern in complying with them, and in the relation between culture, tradition, religion and customs versus the civilized norms that define general progress at the beginning of the twenty-first century.

Notwithstanding our perspective of things, we dare to believe that sustainability is not backed by the status given nowadays to animals. The reasons are mixed and diffuse, with difficulties in distributing the evil produced by natural, institutional, or human causes.

As a general note, at the beginning of the twenty-first century, the poisonous philosophy, with biblical accents, that animals were given to us for sacrifices has not yet been overcome. We still believe we are the masters. We claim to know what lies inside animals' minds and we make efforts to tame them. We know what they are up to and we prevent them from doing such things. We know for certain that they have horns, milk, flesh and ivories, but we do not know for sure whether they possess self-consciousness, feelings and the notion of their own existence. We do not know whether we stress them when we pursue them or we lead them to be killed. We assume that they hurt when we cut their head, but we are not sure whether they are aware that they hurt. All this time, they might love us, of course without knowing that they love us! Let's just say that this is the locus of human hypocrisy par excellence, both for the past and for the present. What can be sadder than the classicists' possible comparison of the "natural condition of slaves" to the "inhumane nature of animals"?! And what can be as sad as the continuation of such convictions?!

In the clumsy relationship between humans and animals, a special kind of "pleasure of living" is also present as an explanatory factor. In the relation to other inert factors, towards which man has not received any biblical command to *master* and sacrifice them efficiently, for reasons of subsistence, the relation with animals received special connotation. All religions have special drawers to establish the shape of this relation.

Even when some animals were deemed sacred and not-to-be-sacrificed, to various degrees, without exception, they all emphasize the superiority of man. It is a superiority of the dominator, who is capable of feelings, with a conscience, a soul, regrets and satisfactions, with realistic and utopic dreams. Man has everything that the adverse part lacks. Thus, the animal must be taken as such, mastered, and given various levels of consideration according to the culture, education, tradition or whims of its *master*. In a world of efficiency, when domesticated animals belong to an agriculture revolutionized by maximum yields, such attention turns into bare indifference. Animals are tortured and suffer, claims a non-economist, the historian Yuval Harari (2015), who is very fond of them. They are locked up in minuscule cages; their horns and tails are mutilated; the cubs are taken away from their mothers; they are selectively bred to create monstrosities. Who cares? The enterprising farmer or transporter who reserves one square meter for three sheep or three goats does not! Efficiency has frozen their souls, if they even have such a thing. But it seems that this "freezing" is a prerequisite, as moral considerations would stall the process if animals were to be treated "properly", not sold as hamburgers "to go"! If we read the standard laws in these matters, included in the European legislation, we can see that they excel in the efficient and humane methods for the animal's end of the road. The animal must be sacrificed without suffering or with minimum pain! What happens before the sacrifice does not raise the same concerns. Up to that point, we grow them "efficiently", castrated and as overfed as possible, helped to enlarge their liver to become fat, tasty and extremely expensive, "productively" stuffed in folds or superposed on transporting ships, on their way to the world that asked for them. Let's go to a world of "pleasure without pain"!

In other words, we are showing signs that we are interested in the world of wildlife. If we can pay them a visit in cages at the zoo, it's even cooler! Anyway, their extinction seems to sadden us. For the remaining ones, we excel in efficiency and their inhumane over-domestication and disfigurement. If they lack self-awareness, they also lack mental problems (it's so pathetic that people can have such thoughts). Why should we worry? A devastating fire that kills 5000 pigs may be announced in the media with a final sentence that sounds like: "Happily, there were no victims". This means, there were no human victims. Burned pigs are no victims. The media breaks the news and moves on! It's not rational to suffer for soulless things!

Unfortunately, the great agricultural revolution is based on such a tearless show. However, how sustainable is such efficiency? It is a productivity issue in tough conflict with a moral one. Everybody's "pleasure of living" may sustain this conflict. It is hard to pass moral judgements in a space also populated with soulless beings. The solitary case of New Zealand, occupied by conscient animals (so says the law in this country), able ones and with a right to subscribe to welfare, may be imitated, but is revoltingly unique. Who is to blame that such an example is not extended? Is it capitalism? Is it socialism? Is it the third world? Anyway, European capitalists come and hunt in the Carpathian forests of the former communist countries. Back home, they have secured the land! In another part of the world, the longest fence that stops the passage of Dingo hounds towards sheep herds, challenged in length only by the Great Wall of China, is not in Somalia! If we think that, as a rule, the person who shoots an animal for its expensive fur or an elephant for its ivory, is not their final user, we might think about market and law, good or bad practices. The market may clarify that unless something is demanded, there is no offer. The law can put up a fight here, including against hypocrisy. The general interdiction of wearing rare furs would be a sign that even the wealthy are able to love animals.

But beyond such dilemmas and the snob questions about the existence or nonexistence of a conscious life or a soul in animals, it should not be hard for the educated human at the beginning of the twenty-first century to understand that animals are part of our life and without them we are difficult to define. The physiocrats have presented them to us in an excessively elevated language, lest we understand anything from their talk. Are we capable today to test whether the animal does or does not have the conscience of its precarity? And what is the use of such a test? It's true that it lacks ultra-specialized "educated" instincts. But it does have healthy and non-perverted primary instincts, possibly more sustainable than ours. We are not talking about smell, taste, speed, etc. Their care for their offspring is mimicable. So is its care for its life partner. The "rational" calculation of the optimal rut moment is related to the generosity (or the lack thereof) with which Nature may welcome the baby. How many mothers in the animal world give birth to their cubs in improper conditions for their normal development? They cannot communicate using articulated language, yet they can say what they feel with their eyes, tail, etc. They know how to show their fear, hunger, cold, or joy. Should the simple fact that they cannot talk convict them to inhumane human action? It's true, it would be preferable that they scream when we stuff their liver or we suffocate them in cattle houses or on ships. They're going to bite us! But they don't, because they lack awareness! Who does Mrs. Brundtland love?

6.7 Concluding Remarks

When the Brundtland Report stated that the environment represents, beside the economy and society, the third important piece of sustainable development, the motivation behind such analysis concerned especially the ecological footprint. It was not bad that an "external authority" gave us a purpose, trying to make us aware of what the statistics had to say. Pollution in all its forms of manifestation and the sad and statistically ascertainable perspective of the finitude of resources were severe warnings for all the administrators of the post-war project of economic growth. To become sustainable, growth must conclude a new contract with Nature.

The text of this chapter is evidence that the signal of the said Report has not fallen on uncharted and unprepared ground, quite the contrary. To various degrees, nature has been a preoccupation and a factor in all the phases that marked the trajectory of the Science of economics. Maybe pollution, the mantra of all people that nowadays experience the paralysing fear of living on a planet that cannot be made longer or wider, has not enjoyed an equally privileged treatment in the works of the founders of this science. But the lack of wariness for a nature that was beginning to be subjected to unfriendly treatment by industries did not concern only economists. A universal relaxation took place on grounds that the wounds inflicted on nature were not yet obvious and sufficiently painful to draw attention. The preoccupation and the scientific treatment of this issue came later. Pigou and Kuznets are the signatories of such beginnings. On these grounds, pollution would turn into an engrossing subject for economic analyses, disproportionately discussed in comparison with other important issues, due to the fact that the anti-pollution noise made, perseveringly repeated, was found to bring guaranteed victory, including rapid scientific recognition. One of the reasons why the consecration of new sciences is attempted, stemmed from Economics and filled with environmental preoccupations regarding pollution, rely on such circumstances. Dealing with the environment sounds good and is in fashion! You should not get entangled with intricate theoretical fields. However, for what Nature represented beyond pollution, for its resources and participation in the agricultural or industrial production processes, its presence was inevitable in the works of the founders. The Earth as well as the other resources have constituted the starting conditions. None of the founders assigned them the mark of infinity. They saw the limitations, but they were not worried. They thought it to be their moral duty to find solutions. Free trade, both internal and external, the control of population and technical progress have reduced their trepidations. First of all, their trust in human intelligence allowed them to be optimistic and confident in the model of civilization based on the free market. They knew the circumstances in which nature may also work for free, when there are resources in excess or conversely, when expenses had to be made. And, in addition, directly or indirectly, it is clear that they support two great ideas: (a) the substitution of nature by anything else was thought of as absurd; (b) technical progress, as much as it was known, was made responsible for the extent to which resources could be known and defined.

Two works have proven to be instrumental for the qualified manner in which the classicists prefaced the future doctrinal tensions about the complementarity or the substitution of production factors. Ricardo's *On Machinery* and Marx's *Fragment on Machines* are two analyses that incubated the embryos of the future theory of "creative destruction". The theory of sustainable development would stand to gain if it capitalized on them as they deserve. We take the risk to claim, yet with good reasons, that Ricardo was more applied, more technical and more logical in clarifying the nature of the causal relation between man and machine, than Schumpeter. The unbeatable logic of the way in which machines do not "grow" alone but together with people, as well as the non-conflictual sequential order in which an old machine is replaced by a new one needed no more additional theoretical continuations. Ricardo's machines are discovered neither "suddenly" nor at a global scale. Each one of them awaits its turn. Meanwhile, a saving is transformed into an investment. This is man's business! Ricardo's "destruction" is neither instantaneous nor happening to everybody at the same time. Its object is not a malignant elimination of labour; the sole discomfort caused to man is not its layoff, but rather the completion and the change from one form of usage into another. But this means adaptation to something new, not the elimination of people from the equation of life. Period! Actually, it is a comma. From Ricardo and Smith, whose ideas he valued, we learn something very important about the role of work in the said equation of life. They correctly saw work as a means that people use to survive. The glorification of work in itself seems to them to be an absurdity as long as they enjoy the fact that the machine simplifies human efforts. If the machine replaces human labour, very well! It can do this without limits, Solow admits. It can do it, but within "dialectical" limits, admit Marx and all the scholars of his line of thought. Work defines man, Engels also says. That means that man needs work, otherwise he cannot be happy. If man does not have it, the government must do its best to provide it for him. It does not matter how the government does this: man must work something. This is the deceiving logic, brilliantly commented upon by M. Friedman when, observing a group of Chinese who were digging a big channel using spades and shovels instead of adequate machines, spurred their leader to give them spoons instead, as their declared objective was a job-creating Plan.

When he started his journey, Schumpeter admittedly intended to grapple not with Ricardo, but rather with Marx. From Marx the "technician", he took the lesson and processed it pretty well. But the temptation to also nourish his ideas from Marx the muddy dialectician was too strong. Thus, between the covers of the same book, *Capitalism, socialism and democracy*, the laborious and reliable explanation of the famous "creative destruction" is encapsulated in the great idea that defines the key in which the work was written: future belongs to socialism. Even if it were bleaker and less efficient, the world would prefer it to capitalism, which is too wonky and restless to manage a relationship as full of conflicts as the one between labour and capital. On the whole, through Schumpeter, Marx's perverted dialectics won. Man kills his own child, the machine, because meanwhile, it has become his foe. That is, capital deserves to be negated. Socialism DOESN'T. It also solves the unsolvable. Who can claim to have heard of unemployment in socialism? Nobody! The state-party found a job for everybody, regardless of whether it was useful or useless. Socialism secured its social support by using this perversion transformed into law. This explains the nostalgia of Eastern workers who had a hard time adapting to the conditions of a free labour market after the 1990s. It was hard for them to understand that work in itself hardly made sense.

On the classical stem of ideas, neoclassical founders graft in a consistent manner. Menger, Marshall, Pigou and Jevons are much more present in the literature of environmental sustainability. And this happens for good reasons. Marshall, fully embracing Ricardian logic, supports the idea of the biunivocal relation between the rhythm of technical progress and the dynamics of new knowledge. Pigou's name is linked to the logic of the principle according to which the polluter pays. Jevons' rebound effect opens bridges to Hotelling and Solow. More such examples can be provided. They lead to the conclusion that the neoclassical founders brought to the table new viewpoints, which are many and important. But they did not claim to have come up with a new science to gather their bundle of ideas. Their offer was produced within the science of economics in which they were "born" and which they found necessary to further develop.

Nothing in the evolution of the two theoretical foundational turning points, classical and neoclassical, announced the radicalism of disparities occurred in the 1970s. Back then, the same world started to be seen with suspiciously different eyes. On the one hand, the promoters of "strong sustainability", animated by Roegen's ghosts, were ready to punish anyone who would have touched nature. On the other hand, there are Solow and the scholars that share his opinion that nature is given to us to make use of it. The former scholars are terrified that production and consumption leave an empty space in the world of resources. The latter scholars believe with conviction that no empty space will be left behind, but rather "capacities of production"; "ability" remains as a legacy for future generations. Roegen's followers urge us to pass on nature as we have received it. Solow believes that this amounts to an unfeasible objective. His own models tell him that every generation must take care of itself. What is to be learned by the theory on sustainable development from these radically different positions in respect to the same unattainable objective? We have tried to answer this question, without claiming to formulate any definitive truth. By and large, we believe that beside the prudence rules formulated by Daly & company, few arguments from the offer of the "strong sustainability scholars" can be retained. The least sustainable is the claimed need for a science that is to explain strong substitution! The studies in this field are of great interest. They may stir doubts on whether or not everything standard science says is sufficient or true. But that's it! The rest is just dread and mistrust in human intelligence. And this does not go well with the canons of a recognized science that has even claimed Nobel awards. A new economic science, created mainly by non-economists, without its own method and having its object "in course of" being defined has very few chances of serving sustainability. From the other side, on the contrary, we find several arguments to believe that the offer shaped by Solow's line of thought comes closer to the spirit of Brundtland. It should not mean indifference towards nature, but rather two-way action; we take some, but we also leave some behind, investing in reproducible capital. In addition, we deal with future generations, but we also do not forget about the needs of the current one. Instead of paralysing fear, there should be ability! Everything is not replaceable, even Solow admits, but future generations will be smart enough to accomplish their life plans. On the whole, hope lies in knowledge.

One more thought. The classical and neoclassical legacy is unpardonably little explored and utilized. With very few exceptions, the two camps work on their own. With respect to Ricardo's "machines" or Marx's "general intellects"—there's nothing of the kind! It's a shame! Actually, the two classical works support, as no other works do, the idea that current civilization is based more on the mind than on material facts. In defining sustainability and resilience, general intellects have the floor. And the delicate and apparently complex issue of the digitally unemployed is still about the general intellects. On the whole, we claim that between the fear of living induced by Roegen and his followers, on the one hand, and the courage based on the creative force of people, inculcated by Solow, Stiglitz, Nordhaus and others, on the other hand, it is not difficult to choose. People love life and know that they will not be born a second time; they only live once.

Nature versus institutional "creative destruction" is another piece of evidence where we find that the origin and healthy dynamics of the same world can be seen with different eyes. The main idea is that a "triad"—Acemoglu, Johnson, Robinson (AJR)—fights against Jared Diamond's excessive geographic determinism. For the history of facts, we find all the works of the mentioned scholars of great interest. For the consolidation of a theory on the role of institutions or nature regarding sustainable development, we cannot claim the same thing. The radicalism with which each of the two camps supports its arguments cannot last. Within the formula of development, it is complementarity rather than substitution of the two factors that has an important role to play. It is true that in certain zones of the globe, geography has argued in a more convincing manner; in other zones, institutions mattered more. But we cannot credit either nature or institutions for such development. It is clear and admissible that both of them matter. If anything of nature's unfairness can be institutionally compensated, such an initiative should be saluted. As for the rest, the discussion on primacy is empty. Geffrey Sachs offers evidence that said dilemma, nature or institutions, is false. And he does so by quoting

Smith to definitively explain that, to fight poverty, the complementarity of geography and institutions is superior to choosing between them.

We are closing this chapter with what we consider that sustainability should amount to in the animal world. We have found that no other domain is more exposed than this when it comes to sustainability. We do not believe that we are entitled to divide people into good and bad with respect to the way they treat animals. However, we confess we are tempted to do just that. It's hard to accept that for too many of us, humanity is sterilized when it comes to beings that cannot talk—the mute part of our entire living world. What is to be said here? No conclusion can bandage a theme so bloodied due to the whole collections of soul sick unscrupulous people that live among us! If efficiency has blinded their reason to an unimaginable extent, this is the place where a criticism of economism taken to extremes that have castrated man's innate humanity is entirely justified. Experience has shown that, unfortunately, even fury may cause boredom. We criticize and we are only left with the criticism. This happens either because decision-makers who misshapenly bring about efficiency, for them and their victims, are too strong or because some twisted and hypocritical logic is usually presented to us in the form of a custom. Under the mantle of a traditional or local, seasonal or festival-related culture, a perverted and criminal practice may subject the collective mind to a gigantic indifference. This is the case when, complying with the protocol, we applaud the exploit of the one who tricked the bull until its breath was spilling his lungs through its nostrils; when we fail to be abstractly or concretely revolted that dogs are skinned alive because only in this way their flesh becomes tastier for the aficionado of inhuman pleasures or when a monkey's brain is served directly from its skull. What do we do with ancient or even modern habits of participating in such cruelties without asking uncomfortable questions?! Such questions may be: How come we test them before putting our own lives at risk? Why do we cause them so much pain to make ourselves feel well? In accounting terms, this is an exchange of equivalences: their unhappiness vs our sustainability! What kind of world is this? Not a sustainable one, for sure. Our common future cannot be made compatible with such forms of indifference towards living beings that cannot protect themselves. The new institutional economics theorizes on the reasoning of transforming informal institutions into formal institutions. Customs and traditions that coagulate majorities are turned into laws. The area we have been talking about is not adequate for such judgements, because we hope that "imbecile" practices, as Veblen used to call them, do not gather majorities, and, in addition, twisted as they are, they do not ask for their own transformation into general rules of behaviour. They do not need laws to represent them, and they do not need laws to allow them fishy interpretations. We need laws that include hard punishments in order to induce normal behaviours.

References

Acemoglu, D., & Robinson, J. A. (2001). A theory of political transitions. *American Economic Review, 91*(4), 938–963.

Acemoglu, D., & Robinson, J. A. (2012). *Why nations fail: The origins of power, prosperity, and poverty*. Profile Books.

Acemoglu, D., Johnson, S., & Robinson, J. A. (2001). The colonial origins of comparative development: An empirical investigation. *American Economic Review, 91*(5), 1369–1401.

Ayres, Robert U., van den Bergh, Jeroen C. J. M., & Gowdy, John M. (1998). *Viewpoint: Weak versus strong sustainability*. (Tinbergen Institute Discussion Paper 98–103/3). Tinbergen Institute Amsterdam and Rotterdam.

Barnett, H. J., & Morse, C. L. (1963). *Scarcity and growth: The economics of natural growth resource availability*. Published for Resources for the Future by Johns Hopkins University Press, Baltimore.

Bentham, J. (1823). *An introduction to the principles of morals and legislation*. Oxford Publishing.

Bloom, D. E., Sachs, J. D., Collier, P., & Udry, C. (1998). Geography, demography, and economic growth in Africa. *Brookings Papers on Economic Activity, 1998*(2), 207–295.

Braudel, F., & Murgescu, B. (2002). *Dinamica Capitalismului*. Corint.

Buchholz, W., Dasgupta, S., & Mitra, T. (2005). Intertemporal equity and Hartwick's rule in an exhaustible resource model. *Scandinavian Journal of Economics, 107*(3), 547–561.

Burns, S. (2018). Human capital and its structure. *The Journal of Private Enterprise, 33*(2), 33–51.

Caldari, K. (2004). Alfred Marshall's idea of progress and sustainable development. *Journal of the History of Economic Thought, 26*(4), 519–536.

Chichilnisky, G. (1996). An axiomatic approach to sustainable development. *Social Choice and Welfare, 13*(2), 231–257.

Chichilnisky, G. (1997). What is sustainable development? *Land Economics, 73*(4), 467–491.

Common, M. S., & Stagl, S. (2005). *Ecological economics: An introduction*. Cambridge Univ. Press.
Cordato, R. (2004). Toward an Austrian theory of environmental economics. *Quarterly Journal of Austrian Economics, 7*(1), 3–16.
Costanza, R., Daly, H. E., & Bartholomew, J. A. (1991). Goals, agenda and policy recommendations for ecological economics. *Ecological Economics: The Science and Management of Sustainability, 3*, 1–21.
Costanza, R., & Daly, H. (1992). Natural and sustainable development. *Revista Conservation Biology, 6*(1).
Costanza, R. (2010). What is ecological economics?. *Yale Insights*, 11 May. https://insights.som.yale.edu/insights/what-is-ecological-economics.
Costanza, R., Cumberland, J. H., Daly, H. E., Goodland, R. J. A., Norgaard, R. B., Kubiszewski, I., & Franco, C. (2015). *An introduction to ecological economics*. CRC Press.
Dale, G. (2017). Sustaining what?: Scarcity, growth and the natural order in the discourse on sustainability 1650–1900. In J. L. Caradonna (Ed.), *Routledge handbook of the history of sustainability*. Routledge, Taylor & Francis Group.
D'Alisa, G., Demaria, F., & Kallis, G. (2014). *Degrowth: A vocabulary for a new era*. Routledge.
Daly, H. E. (1973). *Toward a steady-state economy*. W. H. Freeman.
Daly, H. E. (1987). The economic growth debate: What some economists have learned but many have not. *Journal of Environmental Economics and Management, 14*(4), 323–336.
Daly, H. E. (1990). Toward some operational principles of sustainable development. *Ecological Economics, 2*, 1–6.
Daly, H. E. (1991). *Steady-state economics*. Island Press.
Daly, H. E. (1995). On Wilfred Beckerman's critique of sustainable development. *Environmental Values, 4*(1), 49–55.
Daly, H. E. (1997). Georgescu-Roegen versus Solow/Stiglitz. *Ecological Economics, 22*(3), 261–266.
Daly, H. E. (2008). *Beyond growth: The economics of sustainable development*. Beacon Press.
Daly, H. E., & Farley, J. (2011). *Ecological economics: Principles and applications*. Island press.
Dasgupta, P., & Heal, G. (1974). The optimal depletion of exhaustible resources. *The Review of Economic Studies, 41*, 3–28.
David, P. A. (1985). Clio and the economics of QWERTY. *The American Economic Review, 75*(2), 332–337.
Denis, H. (2016). *Histoire de la pensée économique*. PUF.
Diamond, J. M. (2005). *Collapse: How societies choose to fail or survive*. Viking Penguin.

Diamond, J., & Renfrew, C. (1997). Guns, germs, and steel: The fates of human societies. *Nature, 386*(6623), 339–339.
Douai, A., & Plumecocq, G. (2017). *L'économie écologique*. La Découverte.
Foster, J. B., & Clark, B. (2009). The paradox of wealth: Capitalism and ecological destruction. *Monthly Review, 61*(6), 1–18.
Frey, C. B., & Osborne, M. A. (2013). *The future of employment: How susceptible are jobs to computerisation?* Oxford Martin School, Univ. of Oxford.
Fuente, A., Ciccone, A., & European Commission. (2003). *Human capital in a global and knowledge-based economy: Final report*. Office for Official Publications of the European Communities.
Galbraith, J. K. (2012) Who Are These Economists, Anyway?. In D. B. Papadimitriou, & G. Zezza (Eds.), *Contributions in Stock-flow Modeling*. Levy Institute Advanced Research in Economic Policy. Palgrave Macmillan.
Galor, O., & Moav, O. (2002). Natural selection and the origin of economic growth. *The Quarterly Journal of Economics, 117*(4), 1133–1191.
Goodland, R. J. A., Daly, H. E., & El Serafy, S. (1992). *Population, technology, and lifestyle: The transition to sustainability*. Island.
Grossman, G. M., & Krueger, A. B. (1993). Pollution and growth: What do we know? In I. Goldin, & L. Winters (Eds.), *The Economics of Sustainable Development*. MIT Press.
Harari, Y. N. (2015). *Sapiens: A brief history of humankind*. Vintage.
Hartwick, J. M. (1977). Intergenerational equity and the investing of rents from exhaustible resources. *The American Economic Review, 67*(5), 972–974.
Hartwick, J. M. (1978). Substitution among exhaustible resources and intergenerational equity. *The Review of Economic Studies, 45*(2), 347–354.
Hartwick, J. M. (1990). Natural resources, national accounting and economic depreciation. *Journal of Public Economics, 43*(3), 291–304.
Hartwick, J. M. (2008). What would solow say? *Journal of Natural Resources Policy Research, 1*(1), 91–96.
Heinberg, R. (2012). *The end of growth: Adapting to our new economic reality*. New Society Publishers.
Hotelling, H. (1931). The economics of exhaustible resources. *Journal of Political Economy, 39*(2), 137–175.
Huntington, S. P. (1998). *The clash of civilizations and remaking of world order*. Touchstone Books.
Jacobs, M. (1995). Sustainable development, capital substitution and economic humility: A response to Beckerman. *Environmental Values, 4*(1), 57–68.
Jevons, W. S. (1866). The coal question; An inquiry concerning the progress of the nation, and the probable exhaustion of our coal-mines. *Fortnightly, 6*(34), 505–507.
Johnson, R. D. (2017). The moral and social problem of scarcity. *Rediscovering Social Economics* (pp. 31–40). Palgrave Macmillan.

Kahn, M. E. (2015). A review of the age of sustainable development by Jeffrey Sachs. *Journal of Economic Literature, 53*(3), 654–666.

Kiker, B. F. (1966). The historical roots of the concept of human capital. *Journal of Political Economy, 74*(5), 481–499.

Krugman, P. (2009). How did economists get it so wrong? *The New York Times Magazine*, 2 September. https://www.nytimes.com/2009/09/06/magazine/06Economic-t.html

Kurz, H. D. (2010). Technical progress, capital accumulation and income distribution in classical economics: Adam Smith, David Ricardo and Karl Marx. *The European Journal of the History of Economic Thought, 17*(5), 1183–1222.

Landes, D. (1999). *The wealth and poverty of nations: Why some are so rich and some so poor*. W. W. Norton & Company.

Lucas, R. (1988). On the mechanics of economic development. *Journal of Monetary Economics, 22*, 3–42.

Malthus, T. R. (1983). *An essay on the principle of population*. Penguin Books.

Marshall, A. (2013). *Principles of economics*. Palgrave Macmillan.

Martinet, V. (2012). *Economic theory and sustainable development: What can we preserve for future generations?* Routledge.

Martinet, V., & Rotillon, G. (2007). Invariance in growth theory and sustainable development. *Journal of Economic Dynamics and Control, 31*(8), 2827–2846.

Martinez-Alier, J. (1990). *Ecological economics: Energy, environment and society*. Blackwell.

Marx, K. (1973). *Grundrisse*. Penguin Books.

Marx, K. (1990). *Capital* (Vol. 1). Penguin Classics.

Mayumi, K. (2001). *The origins of ecological economics: The bioeconomics of Georgescu-Roegen* (Vol. 1). Routledge.

Mayumi, K., Giampietro, M., & Gowdy, J. M. (1998). Georgescu-Roegen/Daly versus Solow/Stiglitz revisited. *Ecological Economics, 27*(2), 115–117.

McKenzie, W. (2018). *General intellects: Twenty-one thinkers for the twenty-first century*. Tracus Arte

Mill, J. S. (2015). *On liberty, utilitarianism, and other essays*. Oxford University Press.

Norwood, F. B., & Lusk, J. L. (2011). *Compassion, by the pound: The economics of farm animal welfare*. Oxford University Press.

Passet, R. (1996). *L'economique et le vivant*. Economica.

Perman, R., Ma, Y., McGilvray, J., & Sydsaeter, K. (2003). *Natural resource and environmental economics* (3a ed.). Pearson Education.

Pribram, K. (1986). *Les fondements de la pensee economique*. Economica.

Rectenwald, M. (2019). *Google Archipelago: The digital gulag and the simulation of freedom*. New English Review Press.

Ricardo, D. (2001). *On the principles of political economy and taxation*. Batoche Books.

Rostow, W. W. (1990). *The stages of economic growth; A noncommunist manifesto*. Cambridge University Press.
Sachs, J. D. (2003). Les institutions n'expliquent pas tout. *Finances & Développement*, 38–41.
Sachs, W., Loske, R., & Linz, M. (1998). *Greening the North: A post-industial blueprint for ecology and equity*. Zed Books.
Scholz, T. (Ed.). (2012). *Digital labor: The internet as playground and factory*. Routledge.
Sen, A. (1988). The concept of development. *Handbook of Development Economics, 1*, 9–26.
Šlaus, I., & Jacobs, G. (2011). Human capital and sustainability. *Sustainability, 3*(1), 95–154.
Smith, A. (1977). *An inquiry into the nature and causes of the wealth of nations*. University of Chicago Press.
Smith, M. H., Desha, C., & Hargroves, K. (2010). *Cents and sustainability: Securing our common future by decoupling economic growth from environmental pressures. "The Natural Edge Project."* Earthscan.
Söderbaum, P. (2000). *Ecological economics: A political economics approach to environment and development*. Earthscan.
Söderbaum, P. (2007). Towards sustainability economics: Principles and values. *Journal of Bioeconomics, 9*, 205–225.
Söderbaum, P. (2011). Sustainability economics as a contested concept. *Ecological Economics, 70*(6), 1019–1020.
Solow, R. (1974a). Intergenerational equity and exhaustible resources. *The Review of Economic Studies, 41*, 29–45.
Solow, R. M. (1974b). The economics of resources or the resources of economics. In *Classic papers in natural resource economics* (pp. 257–276). Palgrave Macmillan.
Solow, R. (1986). On the intergenerational allocation of natural resources. *The Scandinavian Journal of Economics, 88*(1), 141–149.
Solow, R. M. (1997). Georgescu-Roegen versus Solow/Stiglitz. *Ecological Economics, 22*(3), 267–268.
Solow, R. (2000). Sustainability: An economist's perspective. In R. N. Stavins (Ed.), *Economics of the Environment* (4th ed., pp. 505–513). W.W. Norton.
Stiglitz, J. E. (1974). Growth with exhaustible natural resources: Efficient and optimal growth paths. *The Review of Economic Studies, 41*, 123–137.
Stiglitz, J. E. (1997). Georgescu-Roegen versus Solow/Stiglitz. *Ecological Economics, 22*(3), 269–270.
Toman, M. A. (1995). The difficulty in defining sustainability. *Resources, 106*, 3–6.
Vernadsky, V. I. (2002). *La Biosphère*. Éditions du Seuil.

Wackernagel, M., & Rees, W. E. (1996). *Our ecological footprint: Reducing human impact on the Earth*. New Society.

Wallerstein, I., & Collins, R. (2014). *Does capitalism have a future?* Oxford University Press.

Wark, M. K. (2017). *General intellects: Twenty-five thinkers for the twenty-first century*. Verso Books.

Zaccaï, E. (2002). De quelques visions mondiales des limites de l'environnement. *Développement durable et territoires. Économie, géographie, politique, droit, sociologie*, (Dossier 1).

CHAPTER 7

General Conclusions

We endeavoured to find out in the pages of this volume to what extent the generous offer for the theory and politics of sustainable development that is to be found in the works of the founders is also present in the works of modern and contemporary economists. If we expressed our own opinions regarding the topics at hand, and we did so diligently, it can be concluded that the message of our book is a liberal synthesis of the theory of sustainability. Bearing that in mind, in the current volume we have studied both renowned names of contemporary liberalism, while not shying away from those haunted by the shadows of the implacable finitude of the world. Using such a framework for our work, we wondered where does wealth come from. Because, if it exists, it can be democratically redistributed. And, as such, we set out to find the special unequal bearing the name of entrepreneur, an unpopular character but very useful in ensuring the healthy dynamics of economics and without whose contribution, sustainability amounts to no more than a chimera. If social peace belongs to those who are "well fed, dressed and sheltered"—and this idea belongs to the liberal founders—then we can safely say that they cannot be forgotten or avoided. Because, we have to add, the spectacle of the civilized world is one featuring the free market and private property, competition and good practices, that are to be found in liberal capitalism and not in planned socialism. Such is the type of sustainability we set out to find, both to the

I. Pohoaţă et al., *The Sustainable Development Theory: A Critical Approach, Volume 2*, Palgrave Studies in Sustainability, Environment and Macroeconomics, https://doi.org/10.1007/978-3-030-61322-8_7

right and to the left. We showed what we found, concretely, at the end of each chapter. Here we just want to add a few points.

First of all, we were struck by the confusing brilliance of the manner in which today's great economists relegate economic phenomenology, through well-targeted policies, to an area of inaccuracies incompatible with the essence of sustainable development. Thus, for example, we were baffled by the way Mises constantly moves between two worlds. He would like to be a descendant of the classicist and, in this position, he wants monetary stability as the prerequisite of an economic cycle without painful syncopes; he wants free competition, also within the banking sector. He is afraid of the conjunction between the servility of bankers and the demagoguery of politicians. He wants rigorous monetary calculation in an economy that is restricted only by law. On the other hand, adherence to the exclusivist perspective of money as a means of exchange, Menger's mantra, prevents him from quoting the classicists and considering the natural rate of interest as an effective lever for settling the nominal economy in the shadow of the real economy. Hayek? Allergic to the road to serfdom induced by the Marxist scabies and a pure ultraliberal, he gives us splendid lessons on the dynamics of production structures, with prices but without technical progress, and ends with a sigh while prescribing a genuine Keynesian therapeutic recipe. Besides, he could not have ended up in another place; like everyone else who thought and wrote like Menger about money! The question is whether central bank governors think like Menger and not like Ricardo or Wicksell! What do we do then? Which of them do we think of when Greenspan calls the fire department after having started the fire himself? Stiglitz accuses the IMF and the World Bank of pushing market fundamentalism; we are warned to pay attention to the cunning slips of self-interest and guiding us towards the state. A friendly state, a repository of collective wisdom tasked with achieving full employment and the promotion of innovation. Where should we quote the classicists? For them, full employment is being taken care of by the business environment. Minsky moves within the same framework when he invites the state to create jobs *directly*. Scholarly perversions? And without a right of reply precisely because they are scholarly? Things that seemed simple to the founders become very difficult to understand if we want to understand them from those who take centre stage today. True, these are just some brilliantly rediscovered things, but what is the use if, at the end of the road, their analyses are disarming or they send us to a global socialism, with an administrator that collects taxes on everything. We do not know if, as a freshwater or saltwater economist, Krugman (the classification belongs to him) worships Piketty for his *Capital in the twenty-first century*. But in doing so, he embraces the latter's belief of an equalizing world state. What about Solow for whom, we have seen, Piketty *is all right*?! It is known that economic dynamics does not occur under the direct guidance of theory but of the doctrine streaming from theory.

It is disarming to see how excellent positive analyses, belonging to some patriarchs of economics, do not find a natural, faithful extension within the normative domain; or they are left to be wrapped up in wandering fantasies! We saw a ray of hope in Keynes, as we announced in a title. Like the renowned mentor of the Austrians, he saw in the standard of value an obstacle to the "green cheese" production. But his path towards rational and inspiring analyses is not clogged. Wicksell's ideational vicinity placed his judgement between definite landmarks bearing a corrective role. For him, the production of money stops when the tension between the natural rate and the bank loan pulls the signal. The liquidity trap operates within this logic. It is baffling to see all the "crisists" quoting from the *General Theory* when looking for arguments for nebulous "relaxations", all the while forgetting his capacity of acrobat in handling the natural rate in the *Treatise on Money*. Just as reasonable is also the kind of prudence promoted by Friedman's school. Maintaining a relationship of proportionality between the money supply and the GDP dynamics is an idea that brings the scholar closest to the classical school, among the economists concerned with the issue of the economic cycle. And, if the story in which he urged the Chinese engaged in digging a large sewer, to do this job using spoons to multiply jobs is true, if that's what they set out to do, then he remains one of the most allergic to the perversions to which the sentences of the science he served may be subjected.

Other than that? Other than that, we should not be surprised that economic science is politically manoeuvrable and economists are quoted in order to provide scientifically sounding justifications in areas where we should not find them. Here are just a few sequences of such a thought:

- In a healthy economy, money and banks must serve; not the other way around. If at the end of the journey among great economists we find out that the nominal economy "is milking" the real economy, the fact also happens because economists have specialized. Those who deal with the movement of money have nothing to do with the movement of goods. The founders did not do this. The legend of King Midas was known to them. Who is still afraid today of the truth it contained?

- Money also gives content and form, anatomy, and physiology of two major macroeconomic policies: monetary and fiscal. For didactic reasons, their theoretical disjunction and autonomy is possible; practically it is illogical. When this decompaction takes place in practice, it is a sign of ignorance and arrogance full of risks. The founders did not separate them. Contemporaries do it and sustainability stands to lose.
- Employment was a manifest concern present in all the founding works. The "golden triangle" designated full employment as the third piece necessary for achieving equilibrium, along with growth and a constant level of prices. Apart from Keynes, who else integrates and analyses them as links in an integrated system? Few or very few. Among those analysed, not even Hayek, in his splendid and scholarly lessons, does not want to be bothered by employment. "Specializations" on the money cycle or "financial engineering" break away from the organics of the ensemble and turn employment into a Cinderella. It is also worth noting that employment has been a heavy, indispensable part of social peace. However, if for the founders the idea above is a constant of their preoccupations, most of the contemporaries remember such a thing only when needed, in times of crisis. And then, it subdues economic logic to social imperatives that, exacerbated by a lack of permanent concern, suffocate or pervert economic logic.
- Inflation and unemployment have proved to be unwanted but permanent companions of the modern economy; one more hostile than the other. Being equally harmful and corrosive, it is inappropriate to compare them. And yet, economists are proposing modern policies to fight inflation, as a means, against unemployment. And in this fight, they define inflation according to the circumstances! For Friedman's school, inflation is, par excellence, a purely monetary phenomenon. Is that so? The Austrians do not like inflation money created out of thin air either, but they can't find a seat at the same table with Friedman and his supporters. For Friedman, the total mass matters, for Hayek's Austrians it does not. When critically analysing banks, Rothbard hints that inflation is a banking phenomenon. Fisher's "endogenous" borrowing points us towards

the difference between receipts and payments, which also generates inflationary outbursts. Minsky's "financial fragility" points in the same direction. Keynes is not afraid of inflation; he sets its limits and turns it into a means. Instead, Laffer sees it as a real and dangerous hideousness and turns his entire arsenal against it. The conclusion? We do not really know what inflation is, but we empower it and turn it into a purely monetary, banking, etc., phenomenon. Anyway, it's a perverse evil giving headaches to some, or making others thrive!

If theorists do not understand each other, what are those who hold a baton in their hand to do? They interfere as "saviours" when the fire is burning. After all, what is fuelling the fire? More paper: fake receipts, false guarantees, money with selenarian support! In one word, all deceptions, drunkenness and "animal spirits" condensed into false promises and illusions. They burn but they do not burn to the end because they "burn" their masters: the government, the central bank, large corporations. How awkward and echoless is Hayek's belief that once we get here, it's good to let things settle where they belong, naturally! Not to put out the fire! Why not go for such an idea? Because "major" interests lead elsewhere. And because not even Hayek was very convincing either. The total amount of money does not seem to have any consequence on humanity; he does not value price stability or the intrinsic power of money. Understanding the bank's treacherous game, he subjects it to "a greater exercise of sincerity and publicity"! What a disappointment for Rothbard! A double disappointment if we keep in mind that Professor Mises allows himself to consider Fisher's proposal completely "illusory". However, all this playing with money, playing around one's own tails and those of others, becomes dangerous when unemployment escalates. When the masters turn from mere instruments into decisive factors, they decide that the game should take another turn because bankruptcies and unemployment give great pains. And then, we throw money into an economy that is contracting, in order to save actors whose size and power in distorting free competition would upset all the founders.

- As a general project, all policies aim at continuous economic growth; without syncopes, with resilient self-maintenance mechanisms. Only those who dream of the need to stop growth are out of line. Against this common background, policies are given specific names depending on subsequent objectives or those derived from temporary but surmountable dysfunctions of the dynamics. That is, demand-based policies or supply-based policies may be fashionable at certain intervals. Structurally, they seem to be in opposition. The first wants state intervention, investments across the whole range, including in non-productive areas, high taxes and high public spending. The second, on the contrary, wants a free market, minimum intervention, low taxes and low public spending. In essence, and in relation to the founding lesson, the two policies do not deserve their name. They promote growth from different doctrinal perspectives. They aim to fight unemployment, respectively, inflation. But can't you fight both at the same time? Only when stagflation puts the goals together do we remember the unity of the whole, the economist Doctor Quesnay, with the anatomy of the economic system in mind, and the fact that when you apply a prescription to a living organism you must know that there are always side effects?

- No engine starts without an impulse. The economic engine is no exception. Who gives the impulse, when and how, are the questions that unjustifiably differentiate the answers. Friedman's monetarists believe that the monetary impulse is everything. Mises' Austrians are not too far away either. Where does the money go "first"? Mises and Hayek think we should focus on capital goods. Minsky is afraid of investment; he wants high consumption and little investment. In addition, he also believes in the miracle of the state creating jobs directly. At the opposite pole, Hayek believes that only by sending money towards production structures do we create sustainable jobs; if we send the money "first" to those who consume, we win the elections but kiss resilience "goodbye". "Advised" by Wicksell, Keynes sets the issue in other terms. For him, the impulse comes from the entrepreneur. Subsequently, we send money into the market to lower the credit. Unmistakably, his entrepreneur is a company. In the Austrian cycle, which runs autonomously from the movements of the currency, the entrepreneur can also be a bank and speculation is, here, at home. The inverted pyramids describe a phenomenon that is difficult to control and, faced with which, the Austrians are disarmed. Those closest to us today are looking to move to the gold standard and drive the state out of monetary policy. In conclusion, the monetary impulse is designed to support an upward trend of growth.

- The concreteness of a situation or the doctrinal colouring of ideas can give money very different destinations: to reduce an interest rate and to make investment attractive; to help entrepreneurs move to the next stage of production when the deflationary price outlook becomes discouraging; provide a life-saver to those for whom the "masters" have decided they are too big to go bankrupt; to pay a debt that cannot be paid out of income; beat deflation or target inflation; stimulate those with tendencies towards consumption or "directly" create jobs without stimulating any business; etc., etc. What do we see? We see that the "means of exchange", money, can be expanded or contracted, as the governors of the central banks say. We can use it to fight inflation or unemployment. We can change, as needed, the cause with effect until we learn whether the rate of interest is an instrument, with an a priori established size, or an end road for the free market exercise. No matter how we proceed, with a fine or rigorous adjustment by the central bank, with "expensive", "cheap" or "neutral" money, with fixed rates dreaming of the gold standard, or flexible rates, it is certain that at the end of the road we want quality goods. The longer and more indirect the road, the Austrians tell us, and the more patient we are, the more generously we will be rewarded! The analysis on the idea seems logical. However, we did not find in the structure of any policy a column in which the political authority urges patience!
- The diversity of opinions expressed on "specialized" areas of economic dynamics is the sign of a living, breathing building site of a science that refuses to freeze in dogmas. This would be deemed positive on the whole, if the exasperating inventory of partial opinions would lead to the outline of common points on major issues of the economic cycle. What we have tried to highlight in the pages of this volume is that the contemporary repertoire on the subject is dominated by excessive diversity (not to say chaos) and doctrinal fragmentation. The theoretical quarrel over more or less rational expectations, the inflation-unemployment balance, the right to priority of supply or demand, the most appropriate place to get the money "first", the fine or rough adjustment, manoeuvring rate of interest or money supply, the payment of debts or the sale of assets, the independence of banks or their servitude to the interests of the real economy, too much or too little market and, above all, the obstinate avoidance of the role of the natural rate of interest

and the logic of the formation of the natural price for goods, turn the theoretical dispute into a sterile quarrel. Neoclassical contemporary science is considered guilty for the fact that the theoretical product of many schools, trends, notorious economists, etc., does not translate to conclusions that support a policy with chances for sustainability. Any references to the founders as a judgement check exercise are pale or completely absent. The landscape populated by large monopolies and central banks professing their independence to everyone does not suggest that we would need Ricardo or Wicksell. And yet! Even in the new conditions, the theory of the cycle would have much to gain from their analyses.

- Regardless of the extent to which theoretical-doctrinal competition contributes to shaping the world's economic policies, and regardless of the effectiveness with which political marketing will sell its product, already polished by exposure to almost a century of existence, called globalization, reality will impose its paradigm. A reality in which transnational corporations, central, national or supranational banks but also nation states will be key players. We will be increasingly tempted to approach Perroux or Galbraith to understand that the world is a relationship between the dominated and the dominant and abandon in the arms of history those who believed that the health and beauty of this world depend on competition between many quasi-equals? It would be a sad and certainly losing thought. We believe, rather, that a translation of the founding theoretical register to the new conditions would grant more sustainability to the economic and social policies. After all, what would it mean to answer today, in the age of globalization and the overwhelming presence of multinationals, questions like: (1) What kind of competitive market should configure the relationships between monopolies and small and medium enterprises so that, at the end of the day, to be able to talk about natural price?; (2) What remains of Smith's "bargaining on the market" or of the "judgement of an honest man" put forward by the representatives of the School of Salamanca when Oliver Williamson believes that we can pursue our personal interest even by deception? Who are the "honest people", genuine entrepreneurs and impartial spectators? Can transnational corporations play this role? (3) What do we do with the actor called bank? Suffering from too much independence, can it still be disciplined by the natural rate and, in this way, provide more stability and resilience to the economic cycle? We believe that we can answer, theoretically and factually, all the above questions and many others, yet to be asked. We are convinced that the founders help us here as well. They taught us how to live healthy and in harmony, avoiding syncopes and reasons for major social conflict. It is important to always read and re-read them. We would be more homogeneous and useful in a lesson on cycle sustainability, and not only of the cycle.

- Economy is being assailed by the invitation to embrace degrowth, but both the theory and the phenomenon as such are pieces of a texture in which both the social and the environment move, interdependently. Those who specialized in this issue do not give us a new bone to pick. The tendency to reduce the material component of our existence and to increase, accordingly, the share of the immaterial is old. It is true that the trend today has another pace. The main factors responsible for this direction are the ecological footprint and the resource crisis. The solemnity and the displayed gravity of the discourse and the urge to stop or even degrow, with encouragements such as "stop" and "now", are related to these circumstances. The fact that growth must cease in order to remind us that we live on a limited planet holds, if not a kernel of truth, at least reason for reflection. In fact, we share the idea that this is not the place for a clear separation of the pros and cons and the accounting determination of the winners. The invitation to find resources for sustainable living in "localization" and in the "immaterial" and to see the earth, with all that it contains, as a strategic resource of foremost importance whose exploitation deserves more attention, cannot be denied by a rational analysis. Nor can we turn down the idea that the extra efficiency gained through the industrial and artificial intelligence revolutions allow us to pause, to enjoy a temporary stationarity. It is not logical to "trip on" each other and to increase tension by stubbornly pursuing tiring work when we are satiated. If we do not live well yet, yes, then stopping is nonsense. At the same time, the idea that we have reached the last station and that if we do not stop, we push ourselves against the wall is unimplementable. And then, how far do we pursue dematerialization? Do we drive out the material and consume IT intelligence instead of bread and butter? Ignoring the artificial load associated with the word stationarity, which has nurtured the degrowth trend, we say that we consider it reasonable to find the balance between the pros and cons, instead of adhering to a passionate idea. If we cannot deny the importance of warnings regarding the negative effects and limits of the post-war type of growth, we do not share either the paralysing fear or its consequences on our life. In other words, we do not take the catastrophists of degrowth as a suitable raw material to use in the future structure of the world. And we wonder what kind of life we would live and what we would say to ourselves every morning if we followed the path they have shown? And, by the way, would those who draw the path follow it?!

- The social layer of sustainable development is one of the most generous, as a thematic area. The anger over the wealth of some, the anathema on profit or the propensity for uniformity act as forerunners for a lot of subjects, one more interesting than the other.

  When you do not set out to gain glory through the difficult path of abstractions—ones that not everyone understands—you gather data on differences in wealth and social position and sound the death knell on inequality-generating capitalism. Success is guaranteed because supporters of egalitarianism have always formed a majority. And what do you do, concretely? You pick on, for example, the soul of capitalism, profit. You relativize the classic idea that profit drives the world, but it does not drive it equally for everyone. And the costs as well as the gains are relative. Everyone earns according to the efficiency with which they spend their resources and their ability to take risks. We generally forget that economic activity is not a gesture of gratuitousness. It is argued that not only profit is to blame, but also the one who motivates the need for profit—the economist! An individual with a fatally ungrateful and losing mission. He is only good with money and accounting! His speech about entrepreneurship, risk, competition, work discipline, etc., has nothing to do with happiness! Happiness must be separated from profit! This mission could be relegated to another science, Ecological Economics, which, in addition, knows how to better integrate environmental issues! And so, we tick several objectives: we get rid of economists, essentially bad, we send profit back from where it came, we open the gates only for social entrepreneurs, we seek happiness in gardens full of flowers, and in addition, we also get a new science! It would be the invention of the world! If this were achieved, the planet would be cleaner, resources preserved, and people could take longer vacations! These are the scenarios that anti-economists long for. And they do so with loud voices and with the support of all the uniformists in the world. We talked about Piketty and Pikettism in this context. The essence of what he told lies in the belief that it would be good to receive equal shares, no matter how much we contribute. In order to be listened to, he encapsulates his belief in a deceptive opening to the civilized world. After all, by manipulating data and biasing its bibliographic sources, he sends us, delicately and inadvertently, into the rudeness of a world with the effigy of the USSR, which, to him, inspires security and coherence. He is very upset that those who have money do not have solid social beliefs. He thinks they are repulsive and would confiscate their property. This way, social concerns would acquire a valve to release the negative energies of a raging and immoral capitalism! A little authoritarianism would not hurt this great adventure; including at the level of a world government, seizing from all who are reckless enough to step out of line. Those who have not experienced the foul-smelling realities of communism can still believe in the prophetic piety of a world of equals described by Piketty. Those who have experienced these cruel and distorting realities can endure, out of pure curiosity, the pleasure of reading Pikettian texts, running the risk of getting confused when discovering the key in which the new world is to be built, an alternative to industrial globalization! What should we tell those like Piketty? To go to North Korea and stay there until they fully satisfy their curiosity and shout from there without anyone answering back?

- The consistent reference we made to nature as a splash of colour within the theory of sustainability is worthy of its size. It is intended as a serious place for debate on many topics that seem to have little connection to one another. Apparently because, in fact, they are connected by the attempt to answer the following question: what must be done so that nature (resources, air, water, etc.) is enough for as long as possible for all those who consume it, humans and animals.

  We pleaded, in context, for the rights of animals to benefit from a friendlier treatment from those with whom they share the same space. The famous "Monkey Trial" in Dayton, Tennessee, 1925, is a sign of the secular concern to convince people that there is no reason to send animals to the category of unconscious and painless objects for the mere satisfaction of our pleasures. It is a sign, like all the others that followed, that there is no justification for subjugating them ruthlessly and without remorse based on criteria of efficiency and sinister pleasures.

Within the same space, we made room for the critical analysis of the old and new dispute of the substitution, weak or strong, of the production factors. We looked for arguments to convince that the passionate, exclusive approach has no place here either. Retaining the common sense warnings from the supporters of strong substitution, we pleaded for and argue here, in a concluding manner, that the Solow line of thought is what places our existence and intergenerational relationships in an equation of rationality, engaging and motivating in the fight for life.

Within the perimeter of the same thematic area, we dared put forward opinions about the hot topic of the future: technical unemployment caused by the artificial intelligence revolution. And, on this topic, we tried to find if there is any possible connection between the message of the founders and the fear that many IT professionals will be replaced by robots! It is known that, unlike other industrial revolutions, that of artificial intelligence is distinguished by speed and the special segments of the active population that it affects. The tertiary that absorbed up to approximately 80% of the active workforce from the primary and secondary sectors is believed to be saturated. In addition, the movement driven by this latest revolution is no longer from one sector to another; it takes place inside the tertiary sector, within the framework of which only the original creators in the field and those who take risks will stand chances to get new jobs. The economic dematerialization produced by this movement, as well as the speed with which it occurs fuels and supports catastrophic positions for all degrowth supporters. It is clear and visible that the phenomenon is occurring at high velocity. Question: do “machines go up” too fast, that is, do they go up faster than people, leaving behind Ricardo’s argued and logical hypothesis? Instead of seeking a rational answer to this legitimate question, stationarists urge governments to get more involved. Faced with possible social upheavals, calls are made for redistributive policies, aid for adaptation, management of new changes through an increase in authority; including, if possible, slowing down new changes! Is it indeed possible that “machines” “go up” independently and quickly, leaving people behind? Logically, and we think about the Ricardian logic, it is difficult to accept such a lack of coherence. The technique, the new technology, no matter how sophisticated, is "glued" to its creator. The machine goes up with the help of man, not independently. Our belief is that the alarm has a large share of falseness. The situation is used to promote more state intervention. Since slowing down change does not seem to be an achievable goal, it is more convenient to urge the state to set everyone in his place, something that the knowledge society has not been able to do. The state knows! And it does it loudly, filling the economic and social spheres with its well-known specific commotion, evacuating the ideas that shape the substance of the problem. We forget, for example, that the process we referred to means a permanent accumulation of knowledge, it is a continuously cumulative process; it is, if accepted, an up-to-date "ability", updated on the basis of an exercise of symbiotic cohabitation, produced at ever increasing levels, between man and machine. No one goes up faster; no one is left behind. And it is hard to believe that we will see the streets full of IT workers going on strike!

Index

A

Africa, 185–190
Antieconomism, 109, 111–114, 118, 121–123, 125, 129, 131–133, 135–137, 139, 141, 143–145, 147–149, 153, 157. *See also* Economism
Austrian school, 41, 99, 142, 167

B

Bankruptcy, 15, 49, 62, 63, 68, 85, 86, 98, 134
Brundtland Report, v–vii, 36, 111, 156, 163, 166, 169, 176, 188, 246, 255
Business cycle, 12, 17, 22, 23, 28, 42, 45, 46, 50

C

Central Bank, 9, 13, 28, 33–36, 58–64, 67, 69, 81–90, 92–97, 99
China, 109, 110, 116, 129, 172, 188, 190, 202, 254
Classical school, 28, 269
Commercial banks, 18, 33, 60–62, 95
Credit, 6, 8–11, 13, 23, 28, 30, 33–35, 41, 55, 56, 61, 65–67, 84, 91, 95, 121, 126, 129, 138, 174, 175, 199, 259, 272
Crisis, 3, 8, 26, 31, 35–39, 41, 55, 56, 63, 66, 68, 71, 85, 86, 95, 98, 120, 126, 173, 217, 245, 269, 275

D

Deflation, 54–56, 61, 70, 79, 273
Digitalization, 241
Durable growth, 166, 169, 175, 184, 196

I. Pohoaţă et al., *The Sustainable Development Theory: A Critical Approach, Volume 2*, Palgrave Studies in Sustainability, Environment and Macroeconomics, https://doi.org/10.1007/978-3-030-61322-8

E

Ecological economics, 155, 156, 166, 167, 176, 215, 217, 219, 243–249, 251, 276
Ecological footprint, 133, 172, 175, 182, 185, 186, 188, 190, 193, 194, 202, 255, 275
Economic dynamics, ix, 12, 14, 17, 18, 32, 36, 37, 50, 56, 66, 68, 69, 75, 95, 106, 107, 111, 118, 122, 129, 133, 164, 206, 251, 269, 273
Economic growth, 12, 48, 95, 97, 111, 117, 121, 122, 125, 126, 138, 144, 170–172, 185, 255, 271
Economism, 109–114, 131, 135, 141, 155–157, 184, 190, 195, 197, 260
Employment, 13–16, 17, 37, 38, 41, 46, 53, 56, 59, 62–65, 70, 77, 83, 92–94, 130, 189, 208, 209, 268, 270. *See also* Unemployment
Entrepreneur, 11–13, 15, 22, 24–26, 28, 31–33, 47, 49, 69, 77, 86, 88, 89, 94, 96–100, 108, 116, 124, 131–136, 141, 142, 151, 158, 165, 177, 199–202, 211, 267, 272, 276
Environmental economics, 166, 176, 212, 214, 243, 244
Environmental problems, v, 117, 152, 173
Equality, 17, 104, 105, 124, 126, 128, 129, 140, 151, 156, 198

F

Foreign capital, 117
Free market, 3, 12, 46, 58, 68, 76, 79, 82–84, 88, 94–96, 99, 110, 112–114, 117, 134, 150, 152, 156, 167, 191, 195, 199, 200, 210, 211, 217, 218, 225, 252, 256, 267, 272, 273

G

GDP, 91, 93, 95, 98, 99, 113–118, 125–127, 147, 157, 158, 167, 171, 172, 184, 186, 195, 197, 269
Geography, 117, 209, 214, 217, 226–231, 259, 260
Globalization, 70, 116–118, 173, 175, 186, 193, 195, 196, 231
Government, 11, 23, 28, 36, 46, 50, 55, 60, 62–64, 80–86, 90, 96, 98, 115, 128, 132, 146, 147, 153, 163, 195, 198, 200, 221, 241, 257, 271, 277, 278

H

Happiness, vii, 32, 104–106, 116, 127, 128, 158, 171, 186, 276
Hayek, Friedrich August von, 17–41, 45, 48, 56, 61, 99, 128, 244, 270–272
Human capital, 115, 137, 138, 201, 211, 221, 236

I

Inequality, viii, 64, 105, 112, 119, 121–124, 129, 136–138, 140–142, 144, 145, 148, 149, 226, 276. *See also* Equality
Inflation, 3, 11, 14, 16, 17, 50, 51, 53, 54, 57, 59, 64, 65, 70, 78–80, 83, 89, 92–95, 97–99, 166, 225, 250, 270, 271, 273
Institutions, ix, 33, 62, 63, 112, 130, 136, 138, 156, 183, 217, 227–231, 259, 260
Intergenerational equity, 248

K

Keynesianism, 57, 59, 68, 71

L

Labour, vi, viii, ix, 3, 31, 37, 38, 48, 65, 87, 88, 95, 104, 117, 138, 142, 146, 152, 164, 181, 206–209, 211, 228, 232–235, 237, 239–241, 256, 257
Liquidity trap, 15, 17, 41, 53, 67, 87, 269

M

Marginalism, 15, 218, 249
Marx, Karl, 3–6, 30, 31, 40, 56, 79, 105, 124, 131, 132, 138, 139, 142, 143, 151, 152, 154, 165, 180, 181, 195, 207, 232–236, 238, 239, 242, 257
Mill, John Stuart, v, ix, 3, 5, 23, 24, 30, 38, 104, 127, 129, 130, 132, 140, 147, 151, 152, 154, 163–165, 170, 175, 187, 190, 207, 210, 229, 230, 242, 246, 251
Minsky moment, 60, 61, 66, 71, 95
Monetarism, 57–59, 62, 71
Monetary policy, 9, 52, 58, 59, 66, 76, 93, 96, 272
Monetary theory, 20, 21, 24, 48, 57–59, 98
Money, functions of
  means of exchange, 2–8, 12, 18, 31, 34, 35, 78, 80, 87, 273
  measure of value, 3–8, 10, 30

N

Natural capital, 171, 215, 217–219, 224, 225, 243, 248
Natural resources, 206, 213, 219–223, 236, 240
Nature, v–viii, 6, 26, 29, 57, 61, 64, 67, 83, 92, 104, 135, 142, 150, 155, 157, 166, 167, 169, 175, 178, 180, 182, 184, 187, 199, 202, 205–207, 209–212, 215, 218, 222, 224, 229–231, 237, 238, 246, 251, 255, 256, 258, 259, 277
Neoclassicism, 59, 118, 212, 214, 215, 250
New Institutional Economy, 81, 227
North, 117, 170, 175, 178, 184, 186–189, 196, 230, 231

P

Physiocracy, 156, 217
Pikettism, 136–139, 141, 144, 145, 147, 152, 153, 277
Pollution, 192, 200, 201, 217, 250, 255
Population growth, 144, 194, 222, 223, 228
Poverty, ix, 55, 115, 126, 127, 129, 130, 140, 141, 157, 158, 174, 186, 187, 189, 193, 196, 201, 212, 217, 222, 227, 229, 230, 260
Price, ix, 4, 5, 7, 10, 13–19, 21–23, 25, 27, 29–32, 35, 40, 42, 48, 49, 56, 76, 88, 91, 99, 107, 211, 219, 221, 224, 246, 268, 270, 273, 274
Private property, 67, 139, 146, 151, 152, 183, 199, 226, 236, 267
Profit, 13–15, 17, 22, 27, 33, 36–40, 49, 65, 87–89, 96, 98, 111, 113, 118–121, 123, 125, 131–135, 142, 149, 157, 158, 165, 175, 201, 206, 209, 276

R
Rate of interest
  market, 14, 15, 17
  natural, 13–15, 18, 19, 33, 69, 76–78, 86–88, 90–94, 96, 97, 268, 273
Rationality, vi, 60, 76, 85, 88, 90, 123, 139, 187, 191, 195, 221, 278
Redistribution, 113, 120–122, 127, 129–131, 138, 153, 156, 157, 184, 185, 190, 202, 230
Roegen, Nicholas Georgescu, 122, 123, 165, 167, 171, 174, 178, 179, 190–194, 215–218, 220, 224–226, 240, 244, 247, 258, 259

S
Scarcity, 37, 206, 210–214, 216, 217, 219, 240
Smith, Adam, 3, 7, 13, 31, 38, 55, 69–71, 88, 91, 95, 123, 124, 127, 128, 134, 140, 142, 165, 197, 207, 210, 211, 221, 229, 231, 241, 256, 260, 274
Social concerns, 106, 108, 119, 153–155, 277
Social dynamics, ix, 70, 156
Socialism, 8, 65, 68, 71, 104, 109, 113, 131, 135, 145–147, 152, 158, 183, 186, 198, 241, 250, 254, 257, 268
Social peace, 63, 64, 105, 109, 121, 267, 270
State intervention, 63, 67, 121, 213, 221, 272, 279
Stationary state, v, 27, 29, 164, 242, 244, 247
Stiglitz-Sen-Fitoussi Report, 114, 125, 127, 158
Strong sustainability, 226, 242, 258

T
Taxation, 36, 76, 80, 145, 146, 213
Technical progress, 26, 27, 77, 111, 173, 181, 206, 207, 209–211, 214, 222, 224–226, 232, 233, 235, 237–239, 256, 257, 268

U
Unemployment, 9, 12, 16, 37–39, 53, 54, 70, 89, 92, 96, 106, 166, 209, 211, 217, 241, 250, 257, 270–273, 278. *See also* Employment

V
Value, 2–4, 6, 7, 9, 18, 30, 40, 59, 64, 78, 80, 84, 109, 115, 117, 126, 147, 150, 166, 179, 195, 207, 213, 220, 222, 225, 239, 271

W
Weak sustainability, 207, 215, 224, 244, 250
Wealth, ix, 1, 16, 23, 104, 111, 119, 120, 125, 128, 133, 138, 142, 145, 149–151, 156, 167, 180, 183, 196, 206, 209, 212, 229, 230, 239, 267, 276
Welfare, ix, 48, 81, 82, 114, 115, 154, 157, 225, 231, 251, 254
Wellbeing, vii, 97, 104, 110, 111, 114, 116, 125, 197, 206, 209, 251
Wildlife, 253

Printed in the United States
by Baker & Taylor Publisher Services